The Palgrave Macmillan Animal Ethics Series

Series Editors
Andrew Linzey
Oxford Centre for Animal Ethics
Oxford, UK

Clair Linzey
Oxford Centre for Animal Ethics
Oxford, UK

In recent years, there has been a growing interest in the ethics of our treatment of animals. Philosophers have led the way, and now a range of other scholars have followed from historians to social scientists. From being a marginal issue, animals have become an emerging issue in ethics and in multidisciplinary inquiry. This series will explore the challenges that Animal Ethics poses, both conceptually and practically, to traditional understandings of human-animal relations. Specifically, the Series will:

- provide a range of key introductory and advanced texts that map out ethical positions on animals;
- publish pioneering work written by new, as well as accomplished, scholars;
- produce texts from a variety of disciplines that are multidisciplinary in character or have multidisciplinary relevance.

For further information or to submit a proposal for consideration, please contact Amy Invernizzi, amy.invernizzi@palgrave-usa.com.

Leif Brostrom DeVaney

Animal Conservation Ethics and the Population Problem

A Habilitation on Rehabilitation

palgrave
macmillan

Leif Brostrom DeVaney
Independent Researcher
Duluth, MN, USA

ISSN 2634-6672 ISSN 2634-6680 (electronic)
The Palgrave Macmillan Animal Ethics Series
ISBN 978-3-031-91890-2 ISBN 978-3-031-91891-9 (eBook)
https://doi.org/10.1007/978-3-031-91891-9

Cover illustration: ADW Images via Getty

This Palgrave Macmillan imprint is published by the registered company Springer Nature Switzerland AG.
The registered company address is: Gewerbestrasse 11, 6330 Cham, Switzerland

To Eireen Belle

Preface

This book was written with the conviction that there is something askew in the world of animal conservation ethics. I sensed that a piece was missing—an important piece that was being kept out of the dominant narrative. This piece is wildlife rehabilitation, and the dominant narrative of population-centrism is flawed. What follows is my humble attempt to bring reform to how animal conservation is thought of and practiced.

I would like to thank Phil Jenni and Peggy Naylor for their dedication to wildlife and their willingness to be interviewed for this project, and my graduate school advisors, David Bengston and Robert Blair, for their support in initially tackling this thorny issue. I would also like to give a special thanks to my family—especially my mom, Gina—who have helped me through some very difficult times to make this book possible. Finally, I would like to thank all the animals (both human and nonhuman) who have taught me so much about life as a sentient being.

Duluth, MN, USA Leif Brostrom DeVaney

Series Editors' Preface

This is a new book series for a new field of inquiry: Animal Ethics.

In recent years, there has been a growing interest in the ethics of our treatment of animals. Philosophers have led the way, and now a range of other scholars have followed from historians to social scientists. From being a marginal issue, animals have become an emerging issue in ethics and in multidisciplinary inquiry.

In addition, a rethink of the status of animals has been fuelled by a range of scientific investigations which have revealed the complexity of animal sentiency, cognition, and awareness. The ethical implications of this new knowledge have yet to be properly evaluated, but it is becoming clear that the old view that animals are mere things, tools, machines, or commodities cannot be sustained ethically.

But it is not only philosophy and science that are putting animals on the agenda. Increasingly, in Europe and the United States, animals are becoming a political issue as political parties vie for the "green" and "animal" vote. In turn, political scientists are beginning to look again at the history of political thought in relation to animals, and historians are beginning to revisit the political history of animal protection.

As animals grow as an issue of importance, so there have been more collaborative academic ventures leading to conference volumes, special journal issues, indeed new academic animal journals as well. Moreover, we have witnessed the growth of academic courses, as well as university posts, in Animal Ethics, Animal Welfare, Animal Rights, Animal Law, Animals and Philosophy, Human-Animal Studies, Critical Animal Studies, Animals and Society, Animals in Literature, Animals and Religion—tangible signs that a new academic discipline is emerging.

"Animal Ethics" is the new term for the academic exploration of the moral status of the non-human—an exploration that explicitly involves a focus on what we owe animals morally, and which also helps us to understand the influences—social, legal, cultural, religious, and political—that legitimate animal abuse. This series explores the challenges that Animal Ethics poses, both conceptually and practically, to traditional understandings of human–animal relations.

The series is needed for three reasons: (1) to provide the texts that will service the new university courses on animals; (2) to support the increasing number of students studying and academics researching in animal-related fields, and (3) because there is currently no book series that is a focus for multidisciplinary research in the field.

Specifically, the series will

- provide a range of key introductory and advanced texts that map out ethical positions on animals;
- publish pioneering work written by new, as well as accomplished, scholars, and
- produce texts from a variety of disciplines that are multidisciplinary in character or have multidisciplinary relevance.

The new Palgrave Macmillan Series on Animal Ethics is the result of a unique partnership between Palgrave Macmillan and the Ferrater Mora Oxford Centre for Animal Ethics. The series is an integral part of the mission of the Centre to put animals on the intellectual agenda by facilitating academic research and publication. The series is also a natural complement to one of the Centre's other major projects, the *Journal of Animal Ethics*. The Centre is an independent "think tank" for the advancement of progressive thought about animals and is the first Centre of its kind in the world. It aims to demonstrate rigorous intellectual enquiry and the highest standards of scholarship. It strives to be a world-class center of academic excellence in its field.

We invite academics to visit the Centre's website www.oxfordanimalethics.com and to contact us with new book proposals for the series.
Andrew Linzey and Clair Linzey
General Editors

Oxford, UK Andrew Linzey
Oxford, UK Clair Linzey

Competing Interests

The author has no competing interests to declare that are relevant to the content of this manuscript.

Contents

List of Figures

CHAPTER 1

Introduction

Abstract This book is primarily an attempt at bringing into focus the relationship between the disciplines of conservation biology and wildlife rehabilitation (from here on "rehab"). It is both descriptive and prescriptive. The project that supports this work grew out of the author's sense, gradually developed over time as a conservation biology student, that a story of relevance to conservation has not been and is not being sufficiently told in the North American animal conservation community. An attempt is made in this work to tell the most important pieces of this story, while acknowledging the necessary limitations of such an account. The view taken is largely of a conservationist less than satisfied with the current state of conservation and may be interpreted as a call to reform by more explicit inclusion of an ignored, and at times dismissed, set of practices. To make the foundational underpinnings of the view I take clearer from the outset, I have attempted to distill some of the basic beliefs I hold into 10 statements, which I provide below.

Keywords Conservation • Wildlife rehabilitation • Paradigms • Participant observation • Environmental philosophy

L. B. DeVaney, *Animal Conservation Ethics and the Population Problem*, The Palgrave Macmillan Animal Ethics Series,
https://doi.org/10.1007/978-3-031-91891-9_1

This book is primarily an attempt at bringing into focus the relationship between the disciplines of conservation biology and wildlife rehabilitation (from here on "rehab"). It is both descriptive and prescriptive. The project that supports this work grew out of the author's sense, gradually developed over time as a conservation biology student, that a story of relevance to conservation has not been and is not being sufficiently told in the North American animal conservation community. An attempt is made in this work to tell the most important pieces of this story, while acknowledging the necessary limitations of such an account. The view taken is largely of a conservationist less than satisfied with the current state of conservation and may be interpreted as a call to reform by more explicit inclusion of an ignored, and at times dismissed, set of practices. To make the foundational underpinnings of the view I take clearer from the outset, I have attempted to distill some of the basic beliefs I hold into 10 statements, which I provide below.

These beliefs are supported by assumptions and intuitions which are in turn given enhanced credibility through evidence. The intent was to design this project according to the liberal spirit of free inquiry. While speculative to an extent, a sincere attempt is made to fortify hypothesizing with empiricism. This is all to entertain the real possibility that a metaphysical wrong turn, as it were, has been taken en route to accomplishing conservation goals.

It is important to note that the story to which I refer may be untold not only because it has been ignored or dismissed, but also because it has yet to fully unfold. To the extent that it has not yet unfolded, the account given is a vision of the future. However, there is also a sense in which the conservation story and the wildlife rehab story have been unfolding from different points of origin that have not been in dialogue before. This book tries to take advantage of a chance to bring them into better communication. While conservation biology has largely grown out of the basic research-oriented field of ecology, wildlife rehab takes as its point of departure the applied veterinary medical field. Because of the newness of both wildlife rehab and conservation biology and the understandable focus of each on the development of distinctive research programs, the growth of these fields has gone largely unheeded by the other. However, the disciplines seem to be on a path of convergence rather than divergence, and some adjustments must be made to make this convergence more of a friendly meeting and less of a conflict.

This work does not remain solely at the disinterested level of an evaluation of compatibility between two paradigms, however. To provide a realistic assessment of these special disciplines and their effects on the entities they claim to serve, ethics, and therefore philosophy, must be involved. These ethical ideals, moreover, must be connected to on-the-ground practical concerns, namely, how conservation and rehab are practiced. Theory and practice must be made to meet, and facilitating this meeting is a challenging task. To accomplish this task, several approaches are used. The broader story told in this work is constructed from varied writing and heuristic styles, using different methods. The first part of this book is theory based, focusing on conservation biology, and draws on metaphysical and ethical reasoning, case studies from the literature, and an original thought experiment. The second part is more practical in theme, focusing on wildlife rehab, and introducing personal experiences structured through the social science method of participant observation.

Since it is directed toward broad trends displayed at the disciplinary level, this project is limited in the level of detail it can attain. An attempt is made, then, to stay within topical limits germane to both disciplines. For example, the bulk of the discussion of conservation and conservation biology in this book pertains specifically to vertebrate animals, which are only a subset of the types of entities that are of conservation concern. Given that wildlife rehabbers mainly treat vertebrate animals and will likely continue to focus on members of this group for the foreseeable future, it would seem wrongheaded to write large amounts on points of complementarity and conflict between plant conservation and rehab. Essential to this balance between specificity and generality is an attempt to center on just what sorts of things we are morally obligated, as well as feasibly able, to conserve.

I have stated that this work is largely ethical in nature and should say a bit more about this up front. Ethics has traditionally been conceived as a branch of philosophy about how to act toward rational entities, and this category has been thought by many key Western philosophers to include only humans. There have been some notable exceptions, including thinkers in deontological, consequentialist, and other philosophical and religious traditions. While contemporary environmental ethicists have made some additional attempts to erode this human-centered position, this erosion has, at times, progressed too far to render the relevant concepts either feasible or convincing. The failure of such holistic arguments is perhaps best exemplified by the concept of intrinsic value, which grounded many early attempts at environmental philosophy but has fallen into disuse lately. Part of the reason for this disuse lies in the portrayal of intrinsic value as a quality that exists in most or all natural objects, while there seems to be little evidence of this existence in many of these entities.[1] For an ethic of nature to not devolve into absurdity, questions must be asked and meaningful distinctions made about what sorts of entities natural objects are. Sadly, some of these important questions have been almost completely avoided by scientists and philosophers of the environment. An example is the basic question of whether any serious category errors have been committed in conservation theory, such that mental constructs are reified as externally existing things.

Much of the work in this book has arisen out of a deeply held belief that conservation, properly conceived and practiced, seeks to heal and never to harm, and that this healing aspect of conservation has not yet been borne out. Included are historical accounts, reviews of recent work, and some possibilities for the future. A lot of ground is covered, and a lot more left uncovered but hopefully pointed toward. An attempt has been made to move from thought to action, to think through problematic assumptions inherent in the language and practice of conservation biology and suggest an alternative way of conceiving of these issues and bringing potential solutions to bear. Although a split seems noted by the division into "argument"[2] and "practice" sections, this is largely artificial. Argument and practice are not so cleanly split, and continually inform one another. This complex relationship, however, does not mean that evaluation is impossible or unnecessary. On the contrary, it is difficult yet possible, and often necessary precisely because of the ease with which disciplinary practices, once established and routinized, can become stagnant and insular.

It is the author's hope that the arguments in this work are open to common-sense interpretation and criticism. An attempt is made to use ordinary language whenever possible. At times, specialized terms are used (and defined as necessary), but often such terms will also be critiqued. While conservation practice may benefit from professionalization, the

[1] Different types of intrinsic value have been proposed (e.g., Sandler 2010). However, the extent to which these are appropriately categorized, or are rather properly expressed in terms of a different form of value (such as inherent value), is debatable.

[2] I choose "argument" over the perhaps more conventional "theory" because of the deconstructive nature of much of this work. The project is as much about questioning and challenging received views as it is about building a new edifice of theory.

observation of a closed language can be misleading and detrimental to the character of conservation as a participatory, and largely voluntary, practice.

To give the reader a basic feel for the general argument, I have provided a brief position statement highlighting the major, foundational themes of the work that follows. Although not exhaustive, the following is an outline of the key values held and displayed in this body of work. Some of the statements are negative, in the hope that departures from the received view may be more clearly shown. The direction given by the content of this outline may be observed throughout the book.

First, my position is focused on wild vertebrate animals and is a departure from the norm of individual-based perspectives in its lack of focus on domesticity. Unlike many otherwise similar positions, however, I do not support wildness as a be-all, end-all value. Rather, I simply treat wildness as the "default" state in which most animals live.

Second, an attempt is made to take a position more realistic than ecological holism, and which admits the impossibility of preserving everything in nature. Resources for conservation must be used wisely. However, I also argue against the error of treating all natural entities as ethically symmetrical, and the treatment of the individual organism as just a cog in a machine (as "ecoholistic" perspectives taken to their logical conclusion do).

Third, my position is focused on providing proper care for experiencers of life when this care is necessary. Because of this, it is based on a sympathetic (and, to the extent possible, empathetic) viewpoint. I aim to present this viewpoint in ways that characterize it as different—albeit with certain similarities—from received environmental and animal ethical perspectives.

Fourth, my position is broadly "lifeist" and acknowledges life as a worthy basis for ethical concern. Because of this, I admit the importance of plants and "lower" animals. However, I also argue that those entities who can experience pain are more appropriate recipients of care than those that show little or no awareness or capacity for experiencing happiness or suffering.

Fifth, my view is unique in its positivity. While many animal- and environmental-ethical perspectives are variations on the theme of "do no harm," mine emphasizes positive actions that can be taken. Because of this, it can be placed in the same category as ecological restoration and rewilding efforts.

Sixth, my position takes seriously the claim that humans are part of nature, not apart from it. However, I point out that we possess certain abilities and tools that can be used for the benefit or harm of other possessors of life, and that we have a moral mandate to use these wisely. With great power comes great responsibility.

Seventh, while my position is methodologically pluralistic, I am not arguing that all conservation biologists should be wildlife rehabbers, or that wildlife rehab is the best conservation method or fitting for all situations. What I am essentially arguing for is the inclusion of another set of practices in the conservationist's toolkit that has often been ignored or maligned by scientists, but which will likely become more necessary in today's increasingly fragmented and technologically dense world.

Eighth, my position is indebted to many aspects of the classic environmental ethic, Aldo Leopold's land ethic, but cautions against the extent of the ethical symmetry Leopold espouses. While respect for the land community is essential, it is also important to stress that some pieces of this community merit more respect, and better treatment, than others. Leopold's own lack of respect for individual lives is a case in point.

Ninth, although predation and competition are real phenomena that are difficult to explain at the organismic level, wildlife rehab is an example of how altruism can evolve from previously non-altruistic relational forms between species. Humans helping wild animals is an evolutionarily significant achievement. Therefore, the potential of rehabbing for future studies on interspecific altruism (and/or mutualism) is noteworthy.

Finally, this work represents a contribution to the small body of literature on the philosophy of animal conservation. While the philosophies of biology and ecology are becoming increasingly established as subfields in philosophy of science, along with the normative field of environmental ethics, the contributions of a philosophy of animal conservation largely remain to be seen. The focus on individual wild animals is a further specialization of animal welfare theory.

While the previous statements give a general idea and hint at the foundation of this work, the following chapter summaries give a more detailed description of the work's basic form. Taken with the above 10 statements, these brief summaries should give the reader a sense of the ways in which this book is a unique contribution to the conservation literature. Chapter 1 (the current chapter) is an introduction to the book.

Chapter 2 begins with a seemingly simple question: What does "conservation biology" mean? It proceeds with an analysis of the use of metaphor and the difficulties this poses. "Conservation" and "biology" are both assessed, with an Aristotelian interpretation given. Problems are found with conservation biology conceived as a special type of or area within biology, conservation applied by the field of biology, or conservation applied to the field of biology. It is determined that conservation biology is either not biology proper or is a special type of biology.

Chapter 3 transitions into a treatment of the relevance of scale in the related discipline of ecology. Problems are found regarding evidential support for the existence of "super-organismal natural entities" (SONEs). Of primary interest to conservation is the SONE known as the population. While the concept of the population is an interesting development, severe skepticism exists concerning the real existence of populations and their associated characteristics. It is suggested, contrary to popular scientific belief, that individualists are in fact the true holists.

Chapter 4 takes a step back from the strongly critical nature of the previous chapter and considers an alternative understanding of populations. This is termed the soft view and highlights the possibility of a conceptual emphasis. While the physical unity of populations has been called into question, their existence (or subsistence) as ideas is considered. The argument is made that such a framing is enhanced through statistical thinking and representation. Finally, an acknowledgment of the improvement of the community concept over the population concept writ large is provided.

Chapter 5 progresses into a case study centered on the gray wolf in Minnesota. It is more generally aimed at introducing consideration of the practice of killing as it pertains to conservation and management efforts. A brief history of wolf management is given, with special emphasis placed on the instatement of a Minnesota wolf hunting and trapping season in 2012. The absurdity of focusing on harm that accrues to populations while denying the significance of harm to individuals is discussed.

Chapter 6 marks a transition from a metaphysical to an ethical focus and is an in-depth analysis of the lines of reasoning commonly used to support killing in conservation. Three argument forms are given, focused on nonnative species, native herbivorous species, and native carnivorous species, respectively. Problems with each of these lines of reasoning are pointed out. The influence of environmentally damaging domesticated animals in this decision-making is discussed. Much of the critique centers on the intractable problem of uncertainty regarding harms and benefits.

Chapter 7 is an attempt at beginning to address the troubled relationship between welfare and ecology. Some of the issues treated have been alluded to in previous chapters. This chapter proceeds by investigating six different potential criteria on which a foundational ethic may be based. These criteria are evaluated based on both desirability and feasibility of implementation. The hope is that a firmer footing can be found on which to build future conservation efforts.

Chapter 8 comprises a thought experiment intending to portray a set of related scenarios of conservation relevance. Deemed the "alien encounter," this situation is designed to elicit judgments concerning the ethics of our treatment of animals. Of particular interest here are the issues of nativity and rarity. The scenarios provide a simulation analogous to real-life decisions that would-be wildlife rescuers must face. Finally, agreement is found with deep ecologist Arne Naess' statement that "The negative value of the severe suffering of an animal belonging to a large population has a no less negative value than that of an animal of a small population!" (Naess and Mysterud 1987: 29)

Chapter 9 moves from theory into a more sustained practical discussion. This chapter includes some of the history of wildlife rehab, most notably found in the work of Albert Schweitzer. A comparison is made between Schweitzer and his venerable contemporary, Aldo Leopold. It then proceeds into an account of the inception of the two major rehab organizations, the International Wildlife Rehabilitation Council (IWRC) and the National Wildlife Rehabilitation Association (NWRA). Basic rehab permit categories are discussed, as are criteria for evidence of an emergency in an adult animal.

Chapter 10 presents a picture of two leading Minnesota rehab centers by way of in-depth interviews. The first, Wildwoods, is a general-level facility in St Louis County. The second, The Wildlife Rehabilitation Center of Minnesota (WRC), is a master-level facility in Ramsey County. While Wildwoods' permit does not allow for treatment of endangered or threatened species, highlighting the value placed on common animals, the WRC treats members of both common and endangered species. Despite being very different centers, Wildwoods and the WRC communicate and collaborate well.

Chapter 11 describes the method of participant observation (PO) and gives an account of themes derived from use of this method in the field of wildlife rehab. PO is a qualitative social science method for settings of human interaction. Contextual details of wildlife rehab at the WRC are presented with an emphasis on relevant experiential information. The goal is to provide a snapshot of what it is like to practice rehab at a state-of-the-art facility.

The book concludes in Chap. 12 with a discussion of a progression that has been made from theory to practice. This progression is characterized as a movement from metaphysician to physician, from thinker to doer. Establishment of wildlife rehab as a limiting case of conservation is one of the hoped-for outcomes. This is not to discount the importance of large-scale habitat preservation, but to emphasize the need for a more interactive approach in areas of human encroachment. The appendix consists of field notes from two seasons of PO at the WRC.

References

Naess, A., and I. Mysterud. 1987. Philosophy of Wolf Policies I: General Principles and Preliminary Exploration of Selected Norms. *Conservation Biology* 1 (1): 22–34.

Sandler, R. 2010. The Value of Species and the Ethical Foundations of Assisted Colonization. *Conservation Biology* 24 (2): 424–431.

PART I

Conservation Biology: Meaning and Metaphysics

The first part of this book is an attempt at bringing the branch of philosophy known as metaphysics to bear on conservation biology. It is also an analysis of the language commonly used in conservation and ecology. While many previous attempts at posing ethical questions related to conservation and other environmentally relevant actions exist, few such attempts have engaged adequately with the "rock bottom" approach of metaphysics. Rather than scratching the surface, as has been the case in prior attempts, I have chosen to try to dig into tough questions about what it is we are dealing with when we are doing conservation. The conclusions I come to concerning what matters in conservation, given my deeper starting point, differ significantly from currently held views within the conservation community.

Much of what follows presents an engagement with an approximation of what has come to be known in philosophy of biology as the species problem. This problem centers on issues with arriving at a consensus about what a species is or represents.[1] In discussing the import of the species problem for conservation, Rojas distinguishes between typological and evolutionary understandings of species (1992). This distinction gets at the current lack of agreement as to what the species concept represents. I aim, as the book title suggests, to engage in a critique of a scaled-down version of this problem: What may be referred to as the "population problem." Goudge (1955:279) points toward this problem when he asks and

[1] Ghiselin suggests a radical solution to this problem by considering species as individuals (1974). This approach is rejected in this book.

struggles to answer, "What is a population?" This struggle becomes acutely apparent when populations are finally characterized both as ultimate units themselves and as being composed of ultimate units (organisms).

I do not intend my version of the population problem to be understood as a reference to the ways in which both small and large populations are often deemed risky, either for themselves or for their environment, in a range of nature-oriented fields. I am instead interested, first and foremost, in asking more elementary questions about the basic ontological status of populations in general. This is largely out of intellectual curiosity about a mostly ignored philosophical issue. However, it is also out of concern about the practical effects of population fixation in conservation.

Why has the question of what constitutes a population been so badly neglected? Perhaps it is because the existence of real populations is something of a self-evident truth. On the contrary, however, evidence for metaphysical populations seems lacking. If I were to put forward a hypothesis to provoke interest (and probably conflict), it would be the following: Population-talk is the conservationist equivalent to the "emperor's clothes" (D. Bengston, pers. comm.). There is a latent belief, to the effect that there is a certain elusiveness of a coherent population concept and the concept's would-be referent, that has not been voiced to a sufficient extent and with sufficient clarity concerning the other-than-human world. This hypothesis, I argue, is rather simply and straightforwardly testable by experience or, perhaps, lack thereof. The position I arrive at is supported by metaphysical realism about particulars, rejecting both nominalism and conceptualism as insufficient for treating actual entities (Audi 1999). I will proceed to this position by first examining the meaning of conservation biology, then delving into a review of the entities conservationists claim to treat.

CHAPTER 2

The Meaning of Conservation Biology

Abstract The aim of this book is best served by beginning with a seemingly simple question: What is conservation biology? I do not pretend to be the first to have asked this question; however, I also do not think it is asked often enough by those who consider themselves conservation biologists. Michael Soule asked it in 1985, not long after conservation biology was formally recognized as a discipline. It seems his answer to this question has been deemed adequate by most, and the question has not been publicly asked much since. This indicates something of a reflective weakness for the discipline, which should be strengthened. Four decades have passed since the question was posed, and such a time span, moreover, should warrant its serious reexamination.

Keywords Conservation biology • Wildlife rehab • Meaning • Ordinary language • Ecology

The aim of this book is best served by beginning with a seemingly simple question: What is conservation biology? I do not pretend to be the first to have asked this question; however, I also do not think it is asked often enough by those who consider themselves conservation biologists. Michael

L. B. DeVaney, *Animal Conservation Ethics and the Population Problem*, The Palgrave Macmillan Animal Ethics Series,
https://doi.org/10.1007/978-3-031-91891-9_2

Soule asked it in 1985, not long after conservation biology was formally recognized as a discipline. It seems his answer to this question has been deemed adequate by most, and the question has not been publicly asked much since. This indicates something of a reflective weakness for the discipline, which should be strengthened. Four decades have passed since the question was posed, and such a time span, moreover, should warrant its serious reexamination.

This question is approachable from at least two different basic angles. One is a descriptive historical approach, facilitated by discussion of the collection of actions, policies, and legislation that have historically been referred to as conservation practices. Depending on the degree of resolution or specificity used, such an approach can be attractive. However, a potential danger of this approach is an overemphasis on what conservation has been, rather than what it is and may become. Trying to answer the "what is" question with a keen eye to the past may render the unsound result, in hindsight, of having effectively answered a "what was" question. Perhaps this is a reason for the "what is" question having been posed early on but not often revisited.

Although key figures and landmark events are important to know for practitioners of any discipline, the possibility of getting mired down in attempts at proving hard to confirm claims looms large, with lengthy regression a potential result. This miring is precisely the sort of thing that should not happen to a discipline of the present and future, as the struggle for persistence of life through time is continually happening but must be anticipated if said discipline is to accomplish its goals. So, while aspects of the historical backgrounds of conservation and wildlife rehab will be acknowledged and at times presented in this book, they will not be its main emphasis.

Another perhaps more analytical approach is to focus on the meaning of the words that are used to compose terms that name our discipline of interest—in this case, the term "conservation biology." Such a course of action seems at once both more feasible and more helpful for understanding what the goal of the field in question is if it would help us identify any unchanging, essence-like properties of this field's objects. To suggest this is not to deny the evolution of language,[1] or the fact that language itself has a complex history, but to point out that much can be learned from

[1] An implication of this critique is that language can be made to evolve faster than the phenomena it refers to.

analyzing labels, and that the label of "conservation biology" has not been adequately analyzed. This lack of application of scrutiny is part of a broader critique of science in general, related to the use of terms as mere markers or placeholders rather than holders of deeper meaning and conceptual content.

This seems to prompt the question: What happens when two or more words that are already in common usage are combined to form a new term? Does this not complicate things? Surely the term will acquire a meaning that is in some ways different from each of its constituent words taken alone, while retaining aspects of the meaning of each. The result is an emergent concept derived from some combination of already established concepts. In the case of such terms as conservation biology, restoration ecology, and wildlife rehab, each word in each term functions as a clue to point in the direction of the new, or newly combined, concept. For words used to describe parts of nature or actions taken thereon, there is presumably a link between the internal reality of the experiencer of the concept the term helps conjure, and an external, referred-to entity or process.

The following project proceeds under the belief that words, while necessarily products of human invention and therefore artificial, can nevertheless fit a situation well or poorly; and that terms about the observable world should, as much as possible, remain faithful to the conventional, common-sense meanings of the words that compose them so as to avoid unnecessary confusion. Words, and especially those that refer to putatively physical entities, do not stand alone, but must be weighed against actual practices. Part of the rationale for arranging this work in argument and then practice sections is so that, after analysis of issues is complete and the practice section read, a judgment can be made as to the fit between the terms and practices in question.

The second chapter is an attempt to seek out an essential, central meaning of conservation biology. By essential meaning, I denote that which cannot be disposed of or done away with without a loss of identity occurring. A promising feature of taking such an investigative approach is that it has the potential to highlight possible future changes in practice, if certain values hold constantly true. Motion, and therefore change, is hard to detect without a fixed vantage point. The question is: Can such a point be found for conservation biology? In the conservation that is being advanced, the basic referent that is to be conserved does not change in any substantial sense, but the means we use to conserve it do. So, by conceiving of the relevant basic qualities of this referent as stable, while acknowledging the variability and evolution of the wider world and taking note of relevant changes and trends in these areas, we may arrive at a better understanding of how past practices have been incomplete, and how current and probable future practices may bring states of affairs nearer to an ideal we desire to achieve. But first we should come to a better grasp of what exactly we ought to conserve.

A more specific version of the previously posed "what is" question, which might prove helpful, is: What would an average person with a basic knowledge of English and little knowledge of environmental issues take conservation biology to mean? Put differently, what would someone without special scientific knowledge be likely to think conservation biology refers to? One way of attempting to answer this question is through social science methods such as surveys or focus groups; another is by looking at the terms through the lens of ordinary language philosophy. Each approach holds promise, but I have chosen the latter for this book.

Ordinary language philosophy is a "loosely structured philosophical movement holding that the significance of concepts … is fixed by linguistic practice" (Audi 1999: 635). This can be juxtaposed with an approach that takes superficial linguistic similarities to be indicative of truth-value. Wittgenstein, for example, spoke of our "bewitchment" by language and how this leads to conceptual error (1953: 47e). A significant portion of this book, then, can be seen as devoted to "unbewitching" those who are conceptually confused by improper linguistic fit. If conservation is to grow in the twenty-first century, these are issues that must be taken seriously by those, both professionals and nonprofessionals, who are shaping the discipline. Science does not exist in a vacuum, and the discourse within it should be open to a broad public, especially if its terms are borrowed from everyday language. If the language of a discipline is confusing or equivocating, this could be a sign that some important terms have been "gerrymandered" from ordinary usage. In a manner analogous to its political origins, this type of disciplinary gerrymandering carves out new boundaries of meaning for a term, and then "lifts" it out of the context in which it was previously found. But this lifting may occur unknown to the broader public, who, quite fairly, expect the term to continue to mean what it has meant for them in their lifetime. Rather than taking the time to choose other, more fitting terms, or crafting new terms to denote something, scientists sometimes hastily incorporate previously non-specialized terms into a specialized lexicon. Such a charge has been made regarding the discipline of ecology (Gobster and Hull 2000), and I will attempt to extend it to specific cases. This privatizing of discourse that is of direct relevance to the public can not only create confusion, but also conflict and distrust.

It is all too easy to argue for such a public language approach, however, while neglecting the trying context of discovery scientists must negotiate. It may seem more fitting to many to choose words already in wide usage and apply them to apparently related phenomena of interest than to craft neologisms. However, there is always the danger that the first, dominant meaning or sense given to a term will be transferred to the new, specialized use of the term. This, I will argue, is the case with certain key words used in ecology and conservation biology. If we come to understand just how these terms are inaccurate or misleading, then our concept, and ultimately practice, of conservation may change.

Making the leap from apparently novel observation to naming and description is a complex cognitive process. This process often involves metaphorical comparisons with past experiences that are thought to bear likeness to the new experience. According to Lakoff and Johnson, "The essence of metaphor is understanding and experiencing one kind of thing in terms of another" (1980: 5). Although this comparative approach to gaining understanding is helpful for making sense of the world, it can have hidden ethical implications, as will be discussed. When metaphor ceases to be an early or middle stage of interpretation, it can become a category mistake, one of critical thought's worst errors. The question of whether the experience of metaphor is caused by synaesthesia, or sense-mixing, has even been posed (Slingerland 2008: 160). These aspects of meaning-making by way of metaphor pose acute problems at the levels of both fact and value.

How a term maps onto its object is largely determined by the user's past experiences, and so is prone to bias and subjectivity. There is no clearly objective way to mentally apprehend the thing-in-itself that is not in some way "contaminated" by experience and attendant theory. Rather than adding to the case for total subjectivism, however, these points serve to strengthen the argument that as much care and concern as possible should be paid to the culturally dominant understanding of key words if their meanings are to be clearly communicated. This can be difficult, even among speakers of the same language within the same country. An example of this issue can be seen in use of the word "coke" in the southern and northern United States. In the North, the word usually means a certain type and brand of soda, whereas in the South it often means soda in general. In addition, it can be slang for a banned drug. As confusing as these distinct uses can prove to be, they all at least refer to an objective physical entity external to the mind and are viewed as such. Therefore, no severe category error appears to have been committed in any of the above uses. When such an error does occur, however, the results can be misleading at best, and tragic at worst.

The stance is taken in this work that words of relevance for everyday life, even if scientific, should mean what they appear to mean and not be misleading. Admittedly, this is a rather vague rule. More specifically, they should not challenge what I will call the conceptual precedence criterion. According to this criterion, a prior meaning of a term in common parlance should be favored over a markedly new application of this term. If the adoption of the new term would lead to blatant equivocation, then the new usage should be eschewed and a different term used for the emergent phenomenon. In the case that this criterion is violated, relevant distinctions are ignored in favor of surface similarities to already familiar concepts. An attempt will be made in this book to inquire as to whether this rule is being broken in some key instances.

Rather than proceeding in the popular way of providing a definition and immediately moving on to tackle the real issues, the constituent words of the terms of interest will be viewed with an eye to how their combination is likely to be interpreted by a non-specialist audience. The relevance of conceptual metaphors and their emphasis on apparent similarities is key for this discussion, since the argumentative force of ecology is derived largely from its emphasis on similarity. What follows, then, is an attempt at approaching some of ecology's and conservation biology's theoretical assumptions with a critical eye to any inconsistencies.

Let us proceed by first examining the possible meanings of the terms involved in the disciplinary name itself: Conservation biology. On the face of it, taken as a discrete term, conservation biology appears to mean at least one of three things: A special type of or area within biology; conservation applied by the field of biology; or conservation applied to the field of biology. Let us examine each of these interpretations, starting with the first. Biology is, at its simplest and most direct, the study of life. According to the Oxford English Dictionary, biology is "the study of living organisms, divided into many specialized fields that cover their morphology, physiology, anatomy, behaviour, origin, and distribution" (2024). Some examples of such special subfields include molecular biology, organismal biology, and population biology. These subfields are mainly focused on different biotic scales. Others, such as zoology and botany, are focused on a class of entities existing in a particular mode of living being (animal or plant). Still other subfields, such as anatomy and physiology, are focused on the study of aspects of a particular biological attribute that causes living and flourishing—in this case, the attributes of structure and function. Basic areas of biological study, then, appear to be classed according to spatial or organizational scale, mode of being, or causal attribute. Is conservation biology one of these subfields or areas within the broader field of biology? Is it a type of biology?

It does not immediately appear that "conservation," by itself, denotes focus at any distinct scale, as do "molecular," "organismal," and "population." Holmes Rolston III, for example, discusses the meaning of conservation as applied to different scales and levels of complexity, from physical concepts like matter and energy, to organisms, to ecosystems (1989). In reference to biological categories of scale, molecular biology can be viewed as biology of or about molecules and population biology as biology of or about populations. However, since conservation can be applied across scales, and to simple entities like matter and energy, conservation biology does not seem to fit into this category. It does not make sense to refer to conservation biology as "biology of conservations" as it makes sense to refer to molecular biology as "biology of molecules," since conservation is not a scale at which aspects of living things can be observed.

Next we come to mode of life. This can be thought of as the strategy a lineage has evolved for living and reproducing and includes causal functions that will be discussed shortly. This category includes the basic taxonomic groups of animals, plants, fungi, etc. Biologists who focus on broad taxonomic groups are labeled differently according to their disciplinary specialization, as, for instance, zoologists, botanists, and entomologists. A brief look at this way of categorizing biological study, however, reveals that even such broad designations are too narrow to encompass conservation. Given that there are conservation biologists who focus on species within different kingdoms with varied modes of life, there does not seem to be a basis for claiming that conservation biology is a type of biology aimed at a particular mode of life. However, the alternative remains that conservation can be applied, more broadly, to life in general. This begs the question, though, of whether life is an attribute that applies generally, as it appears to be commonly thought of, or only within specific individuals.

Perhaps we are not being precise enough if we stop at mode of life, however. We can also "loosen up" mode of life to assess causal attributes. Aristotle famously divided causes into the kinds of efficient, formal, material, and final (1941). We can briefly look at how these kinds of causes apply to the foci of current academic fields considered either part of biology or related to it. Material cause is the matter or substance that things are composed of. Atomic units of material are taken as the subject of physics and chemistry, and if they inhere in or compose biotic entities, are often studied under the disciplinary headings of biophysics and biochemistry. However, even material always found in living things, such as a collection of carbon molecules, does not itself take part in life unless imbued with a certain formal organization and function. Relatively simple, nonliving collections of material can indeed be conserved, but they are distinguished from the types of entities specifically conserved by conservation biologists with labels such as soil and water conservation.

Formal and efficient causes, however, are the forces that give life its special character. The formal or structural attribute is generally held to be the domain of anatomy, whereas the efficient or functional capacity is studied under the heading of physiology. Each of these categories is complementary to the other, and apparently needed for life. Form is expressed through morphology and is a major means by which we discern between species and individuals. It can be thought of as housing the efficient inner workings of the body. Without the full complement of these physiological workings clearly divided into different organs and systems, the body would not flourish and might die prematurely. Conservation biology, however, does not specifically focus on isolation of either of these attributes, since they are difficult to divorce, and each is crucial for life. When attempts are made at such physical splitting, harm or death often results. Anatomy is retained for a time after physiology ceases to function, however, and attempts are sometimes made to preserve or mimic this for medical or educational uses. For example, nature museums have an interest in preserving structure without function through fossils and taxidermy; but this is precisely the opposite of what conservationists hope to practice. They do not want a mere portrait of life, but the real, full quality of being alive.

This brings us to final causes. Although scholars have given many interpretations of the concept of final cause, it is generally agreed to refer to the end or purpose for something. An Aristotelian approach would portray the goal for life forms as flourishing or living well, and this state as being dependent upon proper structure and function. Since when organisms die or go extinct the opportunity to achieve or maintain this state of flourishing dies with them, death, and especially permanent extinction, is an improper end to strive toward, certain though it may be. The emphasis on preserving the opportunity for flourishing comports well with the goals of conservation, for to flourish, the option of being alive (and at the species level, extant) must at least be retained. If the basic conditions for life are not present, such as appears to be the case on most other known planets, or are in danger of being lost permanently, as many environmentalists and conservationists contend is the case with our planet, then these conditions must be preserved and restored where possible. This landscape-scale preservation and restoration of the conditions that make life possible has been a major focus of past conservation efforts and will likely remain so in the future. Moreover, it can be seen as an attempt at putting Leopold's land ethic into practice (1949). The extent to which this sort of conservation is biological is unclear, though, since the conditions for life are not the same as life itself. Much of this type of conservation would appear to fall under the prior categories of soil and water conservation.

At this point, I will posit three possibilities for what biology of final causes might apply to. First is the already mentioned "biology" of the conditions for life's existence, which, while certainly relevant to biology, is probably not a part of biology proper. The second possibility is the biology of life's status as a property of living beings, in the role of supporting the continuation of living for a normal lifespan. Bare life might be viewed as less than flourishing at any point in time, but it is at least a precondition for flourishing to obtain. Third would be the attainment of the optimal, but perhaps hard to attain and maintain, state of flourishing. The second and third options appear more truly biological than the first, since they relate to the workings of living things themselves. However, neither is focused on the study of such beings per se, but primarily on supporting the telos of the final cause itself. This is not to say that nothing can be learned from giving this support, whether it be medical or other. But it does appear to amount to, if biology, biology without the need for study. Could this biology of conservation be final in the sense that it is the last stage in the field of biology, where preservation of the "logic" of life takes precedence over its decoding? In arriving at conservation biology, have we arrived at final biology?

Deciding whether we have in fact arrived at a sort of final biology in conservation biology would require a more sustained historical study and attempt at theory building than is the focus of this book, and is perhaps a project for the future. What can be said preliminarily of this conceptualization is that it has potential to save the notion of conservation biology as a special area of biology. An odd implication of this framing, however, is that most of how this sort of putative biology is accomplished is not strictly biological. For example, in conservation biology the application of protective legal mechanisms, enforcement of rules, and following of ethical principles necessary for goal achievement are quite far removed from the data accrual that is at least the proximal goal of laboratory and field biology.

Another way of thinking about what conservation biology is, is to approach it through the lens of pure or basic versus applied fields. Perhaps I have overanalyzed things, and conservation biology is just applied biology. Rather than being focused on observations from which theory can be derived, it is possible that conservation biology is focused on the application of data and theories that have already been obtained and built. How this application happens, however, is presently unclear. How might basic biology become applied, and for what reasons?

One possibility is that conservation biology is to be applied to living organisms, for the continued study of living organisms. On this view, conservation is a means to the end of biological study; it is applied to living organisms so that the field of biology can continue. This amounts to life being preserved for the sake of biology, conservation applied to the field of biology itself by what methods are available to this end. After all, could biologists really be blamed for proactively building job security by keeping interesting specimens around? I will call this reading the discipline-preserving view.

This interpretation is suggested by the title of the 1987 volume edited by Soule, *Viable Populations for Conservation*. The ambiguity of this notion is troubling, as it is unclear whether the main goal is the existence of viable populations or the practice of conservation. While this distinction may seem subtle and insignificant, if scrutinized it reveals much about a discipline in its formative stages. There is certainly nothing wrong with searching for something to conserve, but the smuggling in of implications concerning the sustainability of the discipline hints at opaque motivational content.

Another possibility is that conservation biology is applied to living organisms, for the sake of the sustained lives and/or lineages of the organisms themselves. This interpretation would seem to comport with a common understanding of biology as referring not to the discipline itself, but to that which members of the discipline study. If this is the case, conservation biology consists of conservation efforts of various kinds applied to the service of life as a good thing in itself.[2] This would amount to life being preserved for its own sake, or for the benefits it automatically provides, and would seem to stem from a less selfish intent than the discipline-preserving version. I will call this the life-preserving view.

It should be noted, however, that the two views outlined above may represent an overly dichotomous reading of the goal of conservation biology. Is the mission of conservation biology really either to preserve living samples for human study, or for the continuation of life itself? It may be that both discipline-preservation and life-preservation are goals of equal or unequal importance for conservation, or that one is merely a happy consequence of the other. The way such goals are structured is not merely academic, moreover, since outcomes depend on them. For instance, if maximal life-preservation was the primary goal, this would set up conditions for discipline-preservation to follow, since biology is the study of living organisms. If discipline-preservation was the primary goal, however, it might be the case that only saving of that life deemed interesting or otherwise useful would be focused on. This would likely mean that fewer lives of well-known or abundant species would be actively preserved by conservationists unless there was some profit to be gained from such preservation. While discipline-preservation may be an accidental but positive side effect of conservation, the primary goal it stems from represents too selfish a motive to be inspiring and is too circular to stand up to scrutiny. The statement that "we are doing conservation so that we can continue to do biology" is thoroughly unsatisfying and would fail to garner popular support for obvious reasons. If life-preservation is the true main goal, the motive seems neither selfish nor circular. However, this begs the question of why conservation itself is not framed more pragmatically and pluralistically, rather than as the sole domain of biology.

Perhaps it is not the goal of conservation biology that is purely biological, but the methods used to achieve this goal. According to such an understanding, the methods of biology are used for the conservation of life and not strictly biology proper. This is largely in accord with Meffe and Carroll's definition of conservation biology as "application of biology to the care and protection of plants and animals to prevent their loss or

[2] However, given that "biology" is defined as denoting a type of study, it would seem less misleading if a term were substituted, such as "biotic," without the disciplinary connotations (e.g., "biotic conservation").

waste" (1997: 10). Such an interpretation, however, entails a questioning of the extent to which application of the methods of biology proper lead to protection of the objects of study. Examples abound where the methods of biology do not further the continuation of the lives of organisms, but actively contribute to and even directly cause the deaths of organisms. Animal experimentation requiring the killing of research subjects, for example, is a common biological practice on insects, fish, mammals, and other taxonomic groups. The case is often made that this type of research is for the "greater good," but to whom or what exactly this good accrues is often unclear.

Even in cases where the type of biological research undertaken or applied is less overtly harmful, it is often not clear how it directly contributes to the care and protection of something. When a field biologist surveys members of a certain species in an area of interest, she is basically a passive observer. The reply might be made that we have lost the distinction between basic and applied, and that the end goal of such research is for conservation. Just how biology is to be applied for conservation, given the knowledge derived from basic biological research, though, remains unclear. For example, would moving from field observation to capturing the observed individuals and studying their anatomies, physiologies, or cellular structure contribute to their care and protection? This seems doubtful.

It is less doubtful, however, that methods from many disciplines that are not biological are used within conservation practice. The restraint practiced by a would-be poacher considering his religious beliefs and moral education, and the construction of a wildlife crossing structure over a busy highway, are more the direct results of thought, social processes, engineering, and construction than biology. Protections granted a tract of unspoiled habitat through a wilderness designation are accomplished more by political and legal means than biological. Even such a hands-on practice as veterinary medicine, dependent on biological background knowledge as it is, considers further gains in knowledge as ancillary to gains in health and well-being. Ironically, it may be the case that the caring practices built on biology, such as veterinary medicine, are the least likely to be currently framed as conservation.

It should be added that those aspects of conservation that are or were more heavily dependent upon biology, may not be dependent upon the continued practice of biology in its current form. For example, while it is likely the case that we would not have knowledge of the rarity or nesting habits of birds of paradise without dedicated biologists, it also may be the case that, now that we have this information, continued focused study of this group would produce marginal knowledge gains. It is likely, moreover, that sustained intensive monitoring of certain species would contribute to added stress and therefore have negative conservation value. So while practice of some form of biology may be (or may have been) necessary for conservation to occur, biology is not sufficient as a means, nor does it appear to be the direct method or suite of methods by which conservation is done.

Perhaps greater clarity can be arrived at by attempting to envision the form some of the different versions of the relationship between conservation and biology can be thought as taking. Problems have been found with conservation biology conceived as a special type of or area within biology, conservation applied by the field of biology, or conservation applied to the field of biology. The most promising framing of "conservation biology as a special type of biology" is as biology aimed at facilitating the expression of final causes for organisms, i.e., as final biology. However, how such biology is done has the paradoxical quality of being largely non-biological, at least in the standard sense of the term. If the methods of biology are not necessary for the doing of a type of biology that is often framed as applied, this brings one to question how the practice of conservation is being accomplished.

It was determined, then, that what it would mean for conservation to be applied by the field of biology, using the methods of biology, remains unclear, as it seems that in some cases the methods of biology are in fact opposed to conservation, and in many others at least do not directly accomplish it. Further problems were found with the interpretation of conservation as being applied to the field of biology itself, unless biology is seen in a non-academic sense as being synonymous with "life" or "biota." If it is true that neither the proper goals nor methods of conservation biology are derived from conventional biology, then what role is left for conventional biology in conservation?

One possibility is to view conservation biology as a sort of stagewise progression that is the opposite of what the order of the constitutive words indicates. Such an ordering would be more clearly shown by, for example, the title of the journal *Biological Conservation*, or, better yet, biology and (then) conservation. What I am referring to is the received view that something resembling biological study must precede conservation efforts, even if this biology only consists of observation, counting, and recording—in other words, data collection. A mostly unchallenged assumption supporting conservation efforts has been that it is necessary to have definitive proof of a low number of living instances of a type of biotic entity before resources are used to this end. Additional assumptions underlying the previous one are that number—or, as it is often referred to, "population size"—is a failsafe indicator of need, and that observational approaches such as point counts render sufficiently accurate approximations of actual abundances. Granting for the time being that these assumptions are valid, or at least that they are held by most conservation biologists, we can posit a picture of conservation biology where biology[3] must precede conservation if conservation is to be most effective.

Such an understanding, however, would seem to imply that the knowledge gained from biology is, while an optimizing asset, neither a sufficient nor necessary condition for conservation of living organisms to occur. A form of conservation—if more scientifically haphazard than the "collect data, analyze, decide, conserve" approach—can occur, and has occurred, with very minimal knowledge of biological details. While biological study augments the knowledge base about what we are trying to conserve and can provide helpful information about preferred habitat, food, causes of mortality, etc., this acquired information does not always have a direct application to conservation itself. Moreover, many indigenous cultures have successfully conserved wildlife for millennia without the aid of any strictly biological practices, prior to those cultures allegedly in possession of this reputedly essential knowledge driving species to endangerment and extinction. The model of "in goes biology, out comes conservation," then, is not a realistic picture of how all, or perhaps any, conservation happens. Conservation biology, therefore, cannot just be applied biology, but is more aptly characterized as the application of methods that preserve or improve the conditions for life's flourishing—which ultimately comes down to aiding the concrete expressions of individual lives. Insofar as biology is helpful for conservation, it is helpful in providing knowledge about the range of conditions that can be expected to facilitate and improve or hinder and repress the expressions of lives.

This brings up the question of whether the continued practice of biology will be useful for advancing conservation to the same extent that it has been, or whether there will come a sort of biological saturation point where conservation can successfully proceed with little new biology. Posed

[3] If counting of living entities is sufficient to be considered biology.

differently, is it likely that the usefulness of new biology for conservation is in any way diminished as more, and more specific, biological questions are definitively answered? Or is the prospect for such lasting knowledge unlikely due to the rapidly changing nature of biotic and abiotic interactions? The idea of such a saturation point being reached has been discussed in the context of certain subfields of biology, such as molecular and cellular biology; however, the extent to which this could occur to an entire field is, and perhaps will remain, unclear.

So, although the possibility of conservation biology being accurately, simply conceived as applied biology is in doubt, an alternative view of biological study as supplemental but not necessary to conservation practice has emerged. Potential topics for future research include an evaluation of those biological subdisciplines and practices that are particularly suited to conservation problems, as well as the future of biology. Can the relationship of biology to conservation be reliably portrayed, such that we may come to a better understanding of how the disciplines interact? Answers to this and many other questions about the link between biology and conservation are pieces of a puzzle that has yet to be assembled. Perhaps some small amount of progress can be made toward such assembly if we more closely inspect the meaning of conservation beginning with basic, pre-biotic categories.

Assuming that the primary goal of conservation biology is the conservation of real, concrete, physical entities and not the subject of biology, it seems likely that the strategy of beginning with the relatively simple and progressing to the more complex will help us achieve clarity. The Merriam-Webster Dictionary defines conservation as "a careful preservation and protection of something" (2024). This is a rather vague definition and will need to be accompanied by some examples. What sorts of entities are often referred to as undergoing conservation? Two examples that quickly come to mind are energy and water. Energy is a basic physical class that, although transformable into matter, is thought to be conserved throughout the known universe, in the sense that it can neither be created nor destroyed permanently. A tension can already be seen between the provided definition and this sort of conservation, since no care is given or needed in the conservation of energy per se.[4] As soon as greater complexity is achieved, however, care must be taken to conserve this complexity if

[4] Although the form energy takes may be actively conserved.

it is desirable and its conservation feasible. Only conservation of the most basic stuffs, then, can truly be said to be entirely passive.

Both water and energy are physical and measurable; for instance, we can measure water in liters, and energy in joules. A significant difference between the two is that water is plainly observable by way of the unaided senses, whereas energy itself is generally not. For our purposes, however, the key fact that is shared by both energy and water is that individual examples of neither exist on a level that is meaningful to humans. It does not make much sense to speak in terms of "a water" or "an energy," but it is possible to measure out amounts of each. We can also identify and name the limits within which there are amounts of either—say, a body of water, or a ray of sunlight. But this is not the same as being able to identify individual cases of either.

Another way of stating this is to say that, metaphysically, neither water nor energy are things, but rather stuffs. If an amount of stuff is removed from anywhere it is found—say, for instance, some water is removed from a lake—the essential identity of both the taken and the taken from is kept. Although vastly different amounts exist in, say, my bucket and the lake, the water in my bucket is no less water, and no more "a water," than the water left in the lake. This is largely because of the lack of differentiation of the molecular parts of water. Even so, if I was able to remove only a molecule of water from the lake, we would still not want to call it "a water." Here tension can be seen between the dictionary definition and our current use of conservation, as conservation is defined as applying to something, and not some stuff. While this point may seem inconsequential at present, it emphasizes an important distinction that is all too often ignored completely. But is it applicable to conservation of more complex items, such as life forms?

If an increase in complexity from the molecular to the cellular level is perceived, a concurrent increase in the ability to individuate is shown. Cells are often multiple and contiguous within tissues and organs of organisms, but single cells can constitute whole microorganisms. While generally not visible to the naked eye, cells are the basic building blocks of organisms. They are not merely structural, however, but are known to be crucial for the performance of various organismic functions, and many cells contain additional functional units known as organelles. When cells work together as tissues, at one level they appear to be things in service of a stuff. Large numbers of cells can be removed from an organism, and yet the organism can continue to flourish. However, as individual molecules can be perceived in water with sufficient magnification, individual cells can be perceived in tissues.

At the biotic level of functionally independent cells such as bacteria, something of a clearer contrast between stuff and thing can be observed. A far greater level of autonomy is reached, especially when motility is achieved. Such tiny living beings are organisms—microorganisms—but lack the organizational complexity found in more traditionally conceived organisms such as those in the groups known as insects, birds, mammals, etc. They do not possess the complex nervous systems of many of their more "highly evolved" relatives and have both a very limited ability to respond to their environment and limited, if any, awareness of themselves or the world around them. Despite this relative lack of complexity, such microbes are clearly individuals. The problem, from an ethical standpoint, is that the marked difference in size between their kinds and humankind renders them difficult to account for and respond to at an individual level.

We can, however, notice many larger, macro-organisms if we pay heed. Those members of the animal kingdom that fall within the broad taxonomic categories of mammals, birds, fish, amphibians, and reptiles, as well as plants, are part of many humans' everyday lives. We and our technologies are, in turn, part of many of their everyday lives. Except for many plants, for which the organismic boundary is often blurry, most of these creatures are clearly individuated. Association with other individuals is at least to some degree voluntary—albeit largely driven by instinct—unlike the contiguity of cells in tissues. There is little question that such macro-organisms are things. Because it is the biotic level that comports best with the scale of the types of beings humans are, the organism seems a fitting stopping point for study. However, such thinking has been challenged from many theoretical angles in the recent past, including evolutionary, genetic, and various ecological theories. One of the central issues approached in this book is the import of theorizing from this ethical standpoint.

In the above section I asked, and tried to answer, some difficult questions about the identity and meaning of conservation biology. Two broad possibilities emerged: Either conservation biology is not biology proper or is a special kind of biology—what we have termed "final biology"—that diverges significantly in its form from conventional biology. We will now progress into a discussion of the meaning of "whole" for conservation biology, and potential problems for certain versions of this concept.

References

Aristotle. 1941. *The Basic Works of Aristotle.* New York: Random House.

Audi, R., ed. 1999. *The Cambridge Dictionary of Philosophy.* 2nd ed. Cambridge: Cambridge University Press.

Gobster, P. H., and R. B. Hull, eds. 2000. *Restoring Nature: Perspectives from the Social Sciences and Humanities.* Washington, DC: Island Press.

Lakoff, G., and M. Johnson. 1980. *Metaphors We Live By.* Chicago: University of Chicago Press.

Leopold, A. 1949. *A Sand County Almanac and Sketches Here and There.* New York: Oxford University Press.

Meffe, G. K., and C. R. Carroll. 1997. *Principles of Conservation Biology.* Sunderland: Sinauer.

Merriam-Webster Dictionary. 2024. Conservation Definition & Meaning—Merriam-Webster. https://www.merriam-webster.com/dictionary/conservation. Accessed 8 December 2024.

Oxford English Dictionary. 2024. Biology—Quick Search Results | Oxford English Dictionary. https://www.oed.com/search/dictionary/?scope=Entries&q=biology. Accessed 8 December 2024.

Rolston, H., III. 1989. Biology without Conservation: An Environmental Misfit and Contradiction in Terms. In *Conservation for the Twenty-first Century*, ed. D. Western and M. C. Pearl. New York: Oxford University Press.

Slingerland, E. G. 2008. *What Science Offers the Humanities: Integrating Body and Culture.* Cambridge: Cambridge University Press.

Soule, M. E., ed. 1987. *Viable Populations for Conservation.* Cambridge: Cambridge University Press.

Wittgenstein, L. 1953. *Philosophical Investigations.* New York: Macmillan.

CHAPTER 3

Large Versus Small Wholes

Abstract When conveying opinions about what is to be conserved, conservation biologists often speak in terms of larger-than-organismic units, namely, populations and species. For example, it may be seen fit to conserve the gray wolf (*Canis lupus*) and "the potential to maintain a significant wolf population" (Mladenoff et al. 1995: 279). It is noteworthy that, in the cases of both populations and species, it would appear, in contrast to the above cases of water and energy, we can point out individual cases of each. It seems to make sense, on the face of it, to speak of "a population" or "a species," and this linguistic convention is widely practiced in conservation biology. What's more, the practice of such a convention seems to align more closely with the definition of conservation as pertaining to something, since it appears we can speak of a particular instance of a population or a species. However, this way of referring to supposedly super-organismal natural entities (let us call them SONEs) may not agree with experience, as it is unclear whether a particular instance of a population or a species can be witnessed. If it turns out to be difficult or impossible to identify SONEs—to know them when we see them, or otherwise experience them—then perhaps their status as mind-independent natural entities may come into question.

L. B. DeVaney, *Animal Conservation Ethics and the Population Problem*, The Palgrave Macmillan Animal Ethics Series,
https://doi.org/10.1007/978-3-031-91891-9_3

Keywords Population • Gray wolf • Organism • Perception • Metaphor

When conveying opinions about what is to be conserved, conservation biologists often speak in terms of larger-than-organismic units, namely, populations and species. For example, it may be seen fit to conserve the gray wolf (*Canis lupus*) and "the potential to maintain a significant wolf population" (Mladenoff et al. 1995: 279). It is noteworthy that, in the cases of both populations and species, it would appear, in contrast to the above cases of water and energy, we can point out individual cases of each. It seems to make sense, on the face of it, to speak of "a population" or "a species," and this linguistic convention is widely practiced in conservation biology. What's more, the practice of such a convention seems to align more closely with the definition of conservation as pertaining to something, since it appears we can speak of a particular instance of a population or a species. However, this way of referring to supposedly super-organismal natural entities (let us call them SONEs) may not agree with experience, as it is unclear whether a particular instance of a population or a species can be witnessed. If it turns out to be difficult or impossible to identify SONEs—to know them when we see them, or otherwise experience them—then perhaps their status as mind-independent natural entities may come into question.

Let us try, then, to test the plausibility of SONEs being apprehensible by way of mental simulation. As an example, think about a flock of geese or other bird that is often seen. This is an example of a relatively small-scale SONE that can be easily observed with the naked eye at one time. The flock flies with a certain form—usually resembling a V shape, if a line is imagined fitting the "points"—that is said to be characteristic of the species. This resemblance does not fool most viewers, however, as it is readily gathered that the apparent "lines" do not compose a letter or shape, but a flock of geese. Only a small amount of focus, then, will reveal the true biotic identity of the group. It might be replied that even a child can see that what we are referring to is not a true V, but that a child may not know the correct biological term for such a group of geese. When one sees the group as something to which the word flock (or any other mass term) applies, then one is seeing it correctly, according to biology.

But where is the flock? How do I know a flock when I see it? I might point to the sky, state that the flock is up there, and that I know it is a flock because I see a group of geese. But I would be hard pressed to identify any concrete observable that denotes the flock, which could not also be said to apply to a single goose—aside from the basic V shape, that is. While the naming of this apparent goose-phenomenon as a flock is deemed correct, whereas the perception of a literal V is deemed incorrect, the identifying features of the V are in fact more significant than those of the flock. We might say that "v-ness" denotes a flock of geese, "ball-ness" denotes a school of fish, etc.

At this point, it seems that a radical shift has been made from the kinds of biology we had previously discussed. Whereas our discussion of biology has been focused on the self-contained homeostatic units known as organisms, and the structure and function provided by varied associations of component parts, it appears that we have now arrived at a treatment of truly separate units. Our subjects are no longer contiguous and forced to associate with one another but are separate and able to direct themselves with a degree of autonomy previously absent. Although such groups as flocks and schools provide benefits to individuals such as mating opportunities, increased vigilance, and decreased likelihood of predation, even within them differences in behaviors admit individuality. It might be said that the previously systematic organization found within organisms has been loosened to make way for an increased level of agency.

This radical shift, however, is not radical enough for many conservation biologists. With the advent of remote sensing technology has come a dramatic increase in the ability to compress visual fields "sensed" from an immense distance. Such spatial distance brings with it both knowledge and distancing of another sort: Distancing from the natural human scale. We can now shrink the entire earth into a neat digital image to be viewed on a computer. If an organismic analogy is drawn, continents appear as organs; organisms are tiny, invisible cells. Our world is no longer the *Umwelt* or self-world of von Uexkull (1920), the spatially limited setting through which one physically moves; it is a representation of the entire earth. With this greatly increased power to represent comes a temptation to re-conceptualize immense aggregations as within actual human purview and control. Systems of organization that are posited of organisms are now also predicated of the entire biosphere, as in Lovelock's Gaia hypothesis (2000). Large numbers of organisms are now seen as microscopic cells, able to be sloughed off the skin of the earth as desired. The question is, does the power of this representation of the earth and its creatures primarily distort or enhance apprehension of reality?

Such a take on our enhanced powers of perception, both downscale and upscale from ourselves, suggests a version of spatial relativism. Is anything truly big or small, or do all things only appear as such in relation to the observer? Is scale real? While these questions are too momentous to address sufficiently in this book, common wisdom would seem to dictate that the world experienced by the normal senses is, at the least, the world of proper concern for humanity. While such an answer by no means sets the self-world of humans on firmly objective footing, it is largely derived out of a pragmatic concern for what is in our power to do. A corollary of such a practical and ethical view is that we may preserve and care for entities both below and above our spatial world if possible and desirable, so long as this care and preservation does not entail harm of those entities we have duties to at our own spatial scale.

From a conventional conservation perspective, it may be argued that a flock of common geese is too small a "unit" to be of conservation concern and is therefore a bad example. In other words, we have not yet arrived at a suitable conservation scale for humans to be concerned with. If this is the case, however, it does not seem to support the view of conservation biology as an activity that is about real things; on the contrary, it would appear to challenge such a view. If, instead of a flock of geese we envision a herd of migrating wildebeest (genus *Connochaetes*), or all the wolves in the state of Minnesota, we hardly have better success in apprehending physical unity.

I use the term "envision" as an acknowledgment of the difficulty anyone would have seeing, with eyes, either of these putative entities in its entirety. This is not a trivial point, as it suggests an empirical issue with our relationship to populations. If the real status of populations is not verifiable by way of the senses, what reasons are we given that they in fact exist? Are the portrayals of technology a valid surrogate for the shortcomings of our sensory abilities?

Despite this metaphysical and epistemological problem, the human desire for greater power and control over nature remains preeminent. We will not scale down, and are reluctant to give up our privileged, newly panoptical view, even for the sake of a firmer grasp on our actual self-worlds. At this point, a tension between power and truth is beginning to become apparent. What we are attempting to conserve may be neither a thing nor a substance, strictly speaking, but whatever it is, it is big and/or very complex.

A common belief in contemporary ecological theory is that SONEs are wholes, and that, as such, they can and should be preserved. Some versions of ecological theory, moreover, seem to promote the idea that SONEs have thing-like qualities. A prime instance of such a view can be found in Allen and Hoekstra's layer cake metaphor (1992). In brief, this metaphor is a conceptual schema for understanding varied levels of ecological organization according to any level of complexity. The "cake" is in the shape of a cone, with the base held to include numerous small-scale entities of varying levels of complexity, progressing to the tip, with a smaller number of large-scale entities. An attempt is made to show that ecological classification, properly conceived, is not dependent upon scale, but organizational level or complexity. According to the given explanation, it is entirely possible for an organism and biome of the same size to exist, although it is more common for biomes to be much larger. Antireductionism is taken to an extreme; the base for large organisms appears to be small organisms, the base for large biomes consists of small biomes, etc. It is unclear to what extent this so-called layer cake is layered, and to what extent the layering is mentally constructed. Just what do the layers represent, after all?

Although there are heuristic advantages to such a neatly unified picture of the ecological world, it is just this unification that should be brought into question. While such a framing correctly draws attention to the organizational attributes of different ecological levels, it does not appropriately distinguish between conceptual characteristics and natural biotic properties. Because of this, it appears unclear on key points, such as the representational import of the cake's layers, the apparent lack of interaction of the various ecological levels, etc. The faithfulness of a representation to its source domain is a serious issue, and particularly so for models of natural systems. The layer cake metaphor shows the power of the human mind to condense the reality of nature into conceivable "chunks," which is an admirable accomplishment. However, such a portrayal leaves some important questions about natural levels of unification, as well as potential implications of key differences between these levels for conservation, unanswered.

An implicit rationale for a highly unified view of ecology is the belief that different ecological levels share certain basic properties in virtue of their systematic behavior (e.g., Miller 1978). A term denoting such properties at a given level, however, may have originated at another ecological level, or even in an entirely different domain. To what extent, it must be asked, do they smoothly carry over? Growth and stability, for example, are often predicated of populations by biologists and ecologists. These properties are purportedly measured or estimated and depicted graphically, and from these measurements and estimations often come predictions of future states of populations.

In standard English, such terms as growth and/or stability are used in reference to entities and phenomena such as organisms themselves, technological objects, and relationships, as a few examples. While these uses by no means exhaust the applications of the above terms, an attempt at evaluating how they might apply to biotic populations is important for gaining an understanding of what, in the ecological sense, they refer to. How did they find their way into the language of ecology, and in regard to SONEs? Perhaps this is properly a question for etymologists; put differently, what qualities do these words likely correspond to? Although it may be understood by ecologists and population biologists that such terms are metaphorical, and therefore not to be understood completely literally, the question of what they refer to seems appropriate. Do stable populations only exist on graphs and in the minds of demographers, or are they part of the real fabric of nature? If it turns out that these terms originally refer to concrete entities but are being stretched to cover reified abstractions, then this could amount to the commitment of a category mistake.

Reification has been equated with the commitment of the fallacy of misplaced concreteness, in which an abstraction is misleadingly portrayed as a physical thing or other concrete entity. This is a central concern with the philosophical treatment of concepts in general. Care must be taken to appropriately fit the language used to construct and describe concepts to whatever is being referred to. If such care is not taken, an immense amount of confusion can result, with attendant practical implications.

The notion of the category mistake was advanced by British philosopher Gilbert Ryle in his *The Concept of Mind* (1949), a groundbreaking work in the school of ordinary language philosophy. According to Blackburn, a category mistake occurs when "things of one kind are presented as if they belong to another" (1994: 58). This definition bears striking similarity to the definition of metaphor, indicating a potential relationship between the concepts. Ryle's first and perhaps most memorable account of a category mistake is in the question "But where is the University?" spoken by a fictitious visitor to Oxford (1949: 16). So, is the mistake that of the questioner, or of those who defined what type of thing Oxford is? A case could be made that the visitor simply misunderstands what most people know is the meaning of "university." Since universities and other institutions are nonliving (though they contain living "parts") and owe their very existences to human intentions, it is presumably up to humans how they should be built, understood, and portrayed. However, the same is less clearly the case with those things that are living, sentient, and/or do not primarily owe their existence to human intent and intervention. The broad conceptual category such beings are often placed in, we call nature.

Eco-feminist author Marti Kheel brings the role of metaphor to bear on environmental ethics, pointing out that metaphors function largely unconsciously and, quoting Mark Johnson, stating that it is "inconceivable to think of moral reasoning without reference to metaphor" (Kheel 2000: 218). Kheel goes on to share her belief that metaphorical reasoning is not an appropriate grounding for an environmental ethic, but that metaphors do factor into our thinking about large-scale ecological entities. What sense, then, can be made of our more literal thought about nature?

Let us assess the ecological terms, in hopes that a clearer understanding of what such terms mean might be reached. Of the putative ecological properties that were raised in the context of populations, growth seems the most basically biotic. What, then, is biological growth? According to Thain and Hickman, growth has "a variety of senses" but "usually involves increase in dry mass of an organism" (2004: 320). It is also stated that growth is "usually regarded as irreversible," apart from the possibility of atrophy in some tissues (Ibid.). This account is echoed by Zs.-Nagy, who states that "All biological individuals display a general tendency of increasing their dry mass content during their whole ontogenesis" (1994: 24). Although six distinct forms of growth are listed and described in the entry, each is described as affecting individual organisms, and none described as affecting SONEs. What would it mean, then, for a population to grow?

For something to grow, an increase of some type must occur. Growth need not involve an increase in size, but, as the above definition states, should involve an increase of mass. If this condition of growth—mass increase—is applied to a population, it immediately appears to be an indicator of growth. The mass of populations appears to increase often. But this does not fully capture what growth implies, in an ecological sense, and is therefore a weak indicator of what ecologists mean when they refer to growth.

For a population to undergo growth in an ecological sense, it is not enough merely for an individual, or even most individuals in the population, to have gains in their dry masses; organisms that were previously not members of the population must somehow be added to the population. This addition can occur, the received view holds, through births or immigration. If five new individuals of a certain species were added to an already established population of 100 individuals, for example, this population would have undergone an increase in mass, thereby meeting such a criterion of "growth as mass increase." But for it to be known that something has had an increase in mass, the boundaries of the thing must be definite; we must be attending to a clearly bounded entity. At what point, it might be asked, does one population become two (or three or four…)? What spatial closeness must the members of a given population maintain for it to retain some sort of unity? Most populations conservationists concern themselves with are not composed of contiguous units, and despite graphic representations depicting them as dots or blobs on a map, populations of vertebrate animals are not as unified as reference to them often suggests. They are basically open systems, and the form of growth—if it is growth at all—that occurs in the context of clearly open systems is apparently very different from the more familiar growth that occurs in the closed systems of individual organisms.

If we make the defining criterion of population growth irreversibility, as the above definition states is characteristic of most growth, this is more problematic still, as gains in population growth are often reversed through deaths and emigration. Perhaps growth is just addition of individual units, at whatever level we choose, to some sort of larger whole. If this is the case, additions of cells in organisms and the addition of organisms to populations would both be forms of growth, as is widely held. It is worth noting that cell divisions occur that add cells but do not involve any increase in mass of the organism, and this is not deemed growth by most biologists, nor is the constant increase and decrease of organismic mass through addition and subtraction of water. With these considerations in mind, what might be the defining traits of a growing population?

Imagine a possible, concrete situation for the sake of clarification: A population of farmed deer artificially confined to a fenced-in area of a known circumference produces offspring, which causes both the number of individuals in the area to increase and the average distance between each individual in the enclosure to decrease. This growth, in the parlance of ecology, is exponential, as it is unlimited by resource use (aside from space) since the deer are being continually fed. The enclosure becomes packed with deer, but the area occupied by the population remains the same. Has this population grown?

The dry biomass inside the fence has undoubtedly increased, as has the number of individuals within it, but the area occupied by the deer remains the same. It is unclear whether we should say that the population has grown, become denser, both, or changed in some other way. What is clear is that a change has occurred within the confines of the fence, and that if an increase in density of deer-material has indeed occurred, this increase has been facilitated by human artifice. In more ways than one, this arrangement is unnatural. Such an artificial increase in putative population density and/or growth, however, points to a key distinction between organisms and populations: Necessary proximity of constituent parts. Although the cells of organisms are often, although not always, contiguous, organisms within populations of vertebrates are rarely so, despite Allen and Hoekstra's statement that "spatial contiguity is used frequently as a principal characteristic of population limits" (1992: 205). While it is the case that contiguity is a natural attribute of many sessile creatures, such as corals, the extent to which such entities are true animals can be questioned, given their lack of adult movement. This characteristic of most adult animals—movement—leads into my next point.

Return to the above thought experiment, but now think of the fence being removed. The stressed and crowded deer run, scattering across the landscape. The running of the deer reveals the motility of most normal adult vertebrates. This motility is another crucial difference between cells and organisms (although, as usual, counterexamples exist). If it was not already clear, it is now evident that the fence was an artificial boundary that caused the collection of deer to become more population-like, in that the individuals were forced into a nearly contiguous group. Analogically, the skin of an organism is like the fence holding in the deer. While the skin is an organ that grows and regenerates with the rest of the organism, however, the fence is an artifact, and not able to adapt to the increasing number of deer within its bounds. Additionally, if a breach in the fence occurs, it allows escape and remains a breach unless repaired by an outside agent. The skin of an organism, however, normally repairs itself within a short time span if broken. So, it appears that the analogy can only be "stretched" so far.

The above example, because of the fencing and feeding, may seem too artificial and bordering on domestication. It is unclear, however, what the key criterion for ecological populations would be, apart from a spatial boundary. Many of the attributes predicated of populations by ecologists, e.g., birth, death, and migration rates, refer to the addition or subtraction of individuals to or from the population, and can thus be explained in terms of organism-level phenomena. Moreover, a concept such as range does not seem to detract from the ambiguity of the population concept, as ranges are widely known to change dramatically and unpredictably. Although a range may be said to grow or expand, it is just as often referred to as following a general trend of shrinking or declining over time. Whether expanding spatial extent or range, or biomass increase, is posited as the defining criterion of population growth, certain difficulties appear to exist, both at the level of matching up this type of growing with the conventional biological understanding of organismic growth, and the regularity and predictability with which it is held to occur in actual populations.

Despite the metaphysical issues that have thus far been raised about SONEs, theories such as metapopulation theory, where discrete subpopulations are portrayed as dots on a landscape, enjoy wide acceptance in the ecology and conservation communities. Metapopulations are perhaps the closest thing to "real" populations in existence, both despite the fact that and because their ranges and relative discreteness are often prescribed by human disturbance. This brings up the following question: Are non-island populations more unified in proportion to the level of human-caused encroachment they face on their territories? If this compression or unification occurs because of technological intervention, then it is a thoroughly unnatural criterion for demarcating populations.

Making sense of a defining feature of populations such as growth is difficult, and this difficulty brings into question other qualities assigned to SONEs. Would we have greater success, for instance, making sense of stability? In common parlance, stability is attributed to chairs, tables, and, more abstractly, relationships. In ecological theory, stability generally refers to the tendency for a population to remain at or close to a certain state for an extended period. More specifically, stability is the "name given to a limit population to which actual populations tend when their mortality and fertility remain constant" (United Nations 1968: vii). However, it is stated in the same document that "Most demographers would agree … that the stable populations which they compute are very seldom found in reality" (Ibid.).

It is difficult to grasp the meaning of stability without discussion of another, more basic concept that has been brought into the vocabulary of ecology: Equilibrium. An ecological equilibrium is often evoked through the metaphor of a ball resting on a hill.[1] This ball, representing a

[1] This metaphor is not confined to ecology, but is also prevalent in physics and engineering theory, where it is arguably less of an abstraction.

population, can be disturbed so that it is dislodged, rolls down the hill into an adjacent valley, and either rolls up another hill to rest at its top, or back down into the valley if it does not have enough momentum. Depending on the width and depth of the point, cup, or dip at which the ball is at rest, the equilibrium can be thought of as more or less stable, i.e., likely to remain at its current level of mortality, fertility, and migration for an extended period. Ideally, the ball just remains still at the bottom of the cup. This stillness depicts stability of the population at equilibrium. The depth of the cup is roughly to the ball what stability is to the population; the greater the depth and closer the fit of width to the ball, the greater the stability. If a sufficiently strong force is applied to the ball, however, it may roll, spin, or bounce out of the cup.

The brief explanation I have just given, although helpful for mental simulation of a conceptual process, does not uncover much about what the overall schema stands for. To what extent does the content of the metaphor used match up with what is the case in nature? It might be pointed out that balls and cups are artificial, and that it seems backwards to frame purportedly natural processes in terms of technological objects. Does it make sense to model the workings of an apparently natural system after the workings of an artificial system—even if only conceptually? This question hits on the relationship between metaphor and reality. Does such a complete symmetry obtain between natural and artifactual objects that it simply does not matter what order the "source" and "target" domains are presented in (Slingerland 2008)? Can anthropogenic objects and systems function to effectively teach us about the workings of natural objects and systems, such that there is total symmetry? If what we are interested in learning is how nature behaves, or what its real entities are like, can we base this knowledge on a conceptual schema dependent upon the interaction between artifacts? These are deep and difficult questions about the relationship between mind, matter, and control of what is out there in the world.

This potential problem related to the correspondence of technology and nature may be averted merely by positing the ball as a round stone, and the cup as a natural depression in the ground. Or we may render the objects within the metaphor more abstract, for example, as a generic sphere in a concavity. This does not go far toward addressing the main issues with this conceptualization, however. What is more to the point for us is the fact that balls, stones, and spheres are all unified objects, whereas it is far less clear that populations are. While SONEs are often shown as dots or shaded-in regions on a diagram or screen, this is obviously a simplification. The degree of fit of the source and target domains is important for the development of an accurate conceptual schema, and although a one-to-one correspondence is too much to ask when dealing with complex and/or abstract notions, there should be a reasonable amount of similarity between the object or phenomenon to be conceptualized and the devices used to encourage this conceptualization. The claim is made, for example, that the source domain is most often concrete, and is intended to help picture the more abstract target domain (Slingerland 2008). A related question that has significant import for conservation practice is: Do populations amount to abstract notions, such that their conceptualization must be aided by more concrete representations? If so, what does this imply about their existence as real, substantial entities, and consequently what can be said about their ethical status?

Before offering answers to these questions, let us examine two commonly used metaphors for comparison. The first is simple and easy to understand: The "teeth" of a saw. In this name for parts of a tool, the comparison is implicitly made between body parts of many animals and a saw's cutting edge. Both are very hard and used to divide materials; the main differences are in the composing substances and means of construction. While the source teeth are naturally composed of enamel, dentin, cementum, and pulp, and have evolved over millions of years through the forces of natural selection (but formed within an organism's lifetime), the technological teeth are composed of metal and have evolved through human design and manipulation for a tiny fraction of this time. Despite these marked differences, both types of objects are referred to, most simply, with an identical string of letters and verbal utterance. Also, both are concrete observables, and so a comparison can be readily made.

A more marked difference can be seen in the analysis of the word "fluid." The source meaning of this word is the material attribute of fluidity. A substance is fluid to the extent that it is readily able to change shape. Here, the term denotes something concrete. Let us then postulate our concrete, fluid material as easy-to-visualize fluid concrete. But fluid has other uses denoting immaterial, abstract objects. For example, fluid or fluidity can also refer to movements, plans, ideas, and the like, which are harder to observe directly by way of the senses. Although the concepts about these different fluids may be similar, there is a marked difference in the characteristics of the domains of reference. While the source fluid can be controlled and conserved in the form of liquid water, for example, by various direct physical means, the conservation or control of the other fluids or fluid-like attributes, if possible, cannot be accomplished in the same way.

The notion of stability seems related to the example of fluidity, in that the concreteness of the source referent is fleetingly and rather deceptively lost. When exactly the move from concrete to abstract is made is not clear, but in both cases meaning seems compromised. Theory about a thing or phenomenon that is difficult or impossible to observe but held to exist at some level is being built upon that which can be readily observed. In Wright's "adaptive landscape," from which the ball and cup examples have been taken, we can see a host of ecological abstractions that have been based on concrete observables (Ruse 1990).

In drawing attention to these examples, I am not implying that those using such linguistic conventions are unable to make the correct distinctions. Many such uses are explicitly described as metaphorical or heuristic, intended for helping us to get our minds around difficult concepts. I am concerned, however, that metaphor may not always give the desired meaning—especially when the entities referred to can be helped or harmed. This point, given the practical nature of conservation, has much bearing on the ways in which actions and policy are taken and made.

One of the possibilities implied by the foregoing analysis of the problems related to populations is that, contrary to common assumptions, SONEs may not be true biotic wholes. Might we need to move down a level, as it were, and consider the possibility that organisms are the whole entities that conservationists ought to be aiming to serve? Although Aitken's (2004) critique of holism as the proper focus of conservation efforts is well taken, I am suggesting that it may not be the case that individualists are not holists. On the contrary, it seems likely, from the perspective of a thing-based ontology, that individualists are in fact the true holists. While populations may be wholes of a certain kind—namely, conceptual—their status as biotic wholes is in question due to lack of concrete evidence. It follows from this lack of evidence that, if conservation biologists are concerned with protection of and care for life, then individual lives should not be considered too insignificant for their concern. However, a prominent contemporary case can help show both how little respect currently exists in the conservation world for even large and charismatic organisms, and how population statuses can count against individual organisms.

References

Aitken, G. 2004. *A New Approach to Conservation: The Importance of the Individual through Wildlife Rehabilitation*. Aldershot: Ashgate Publishing.

Allen, T. F. H., and T. W. Hoekstra. 1992. *Toward a Unified Ecology*. New York: Columbia University Press.

Blackburn, S. 1994. *The Oxford Dictionary of Philosophy*. Oxford: Oxford University Press.

Kheel, M. L. 2000. *An Ecofeminist Critique of Holist Nature Ethics: Attending to Non-human Animals*. Berkeley: Graduate Theological Union.

Lovelock, J. 2000. *Gaia: A New Look at Life on Earth*. 3rd ed. Oxford: Oxford University Press.

Miller, J. G. 1978. *Living Systems*. New York: McGraw-Hill.

Mladenoff, D. J., T. A. Sickley, R. G. Haight, and A. P. Wydeven. 1995. A Regional Landscape Analysis and Prediction of Favorable Gray Wolf Habitat in the Northern Great Lakes Region. *Conservation Biology* 9 (2): 279–294.

Ruse, M. 1990. Are Pictures Really Necessary? The Case of Sewell Wright's 'Adaptive Landscapes'. *PSA: Proceedings of the Biennial Meeting of the Philosophy of Science Association* 2:63–77.

Ryle, G. 1949. *The Concept of Mind*. London: Hutchinson House.

Slingerland, E. G. 2008. *What Science Offers the Humanities: Integrating Body and Culture*. Cambridge: Cambridge University Press.

Thain, M., and M. Hickman. 2004. *Penguin Dictionary of Biology*. 11th ed. London: Penguin Books.

United Nations Department of Economic and Social Affairs. 1968. *The Concept of a Stable Population: Application to the Study of Populations of Countries with Incomplete Demographic Statistics*. New York: United Nations.

Von Uexkull, J. 1920. *Umwelt und Imnenwelt der Tiere*. 2nd ed. Berlin: Springer.

Zs.-Nagy, I. 1994. *The Membrane Hypothesis of Aging*. Boca Raton: CRC Press.

CHAPTER 4

The "Soft" View of Populations

Abstract Although, as the foregoing evaluation suggests, there are problems with the metaphysical conception of SONEs, perhaps there is a different conceptualization that is more appropriate. For example, it may be the case that SONEs can be best understood not as physical things or cohesive systems, but as a politically convenient way of selecting aggregates. On such a reading, it would be understood that when we refer to a population, we are taking certain linguistic liberties to obtain more power, whether this power be of a conceptual or political nature. In this sense, population refers more to a methodological framework, a way of thinking about a group that may not fully "hang together," than to a true unit.

Keywords Population • Statistical method • Explanation • Uncertainty • Community

Although, as the foregoing evaluation suggests, there are problems with the metaphysical conception of SONEs, perhaps there is a different conceptualization that is more appropriate. For example, it may be the case that SONEs can be best understood not as physical things or cohesive systems, but as a politically convenient way of selecting aggregates. On

L. B. DeVaney, *Animal Conservation Ethics and the Population Problem*, The Palgrave Macmillan Animal Ethics Series,
https://doi.org/10.1007/978-3-031-91891-9_4

such a reading, it would be understood that when we refer to a population, we are taking certain linguistic liberties to obtain more power, whether this power be of a conceptual or political nature. In this sense, population refers more to a methodological framework, a way of thinking about a group that may not fully "hang together," than to a true unit.

After a distinction suggested by Ariew (2008), let us call this developing idea of populations the soft conception, and the former metaphysical interpretation the hard conception. While the scientific soundness of populations would likely be strengthened if it turns out that they are unambiguously bounded, unified, and not easily divisible, evidence for expression of these attributes has been shown to be lacking. However, by considering a softer version we may at least be able to gain some insight into what positive functions population thinking might provide. How, then, are non-ontological populations to be understood?

Statistical analysis clearly plays a part in this way of thinking, bolstered by unitized and continuous representations such as data points, fitted lines, Gaussian curves, and the like. However, it seems straightforward that, in conventional statistics, individuals are sampled and grouped to form a representation. At least three significant transformations can be seen in this process of representation: From physical entity to experience; from experience to datum; and from datum to statistical knowledge. Many more specific steps could be taken, but these three are sufficient for present purposes. An explicitly statistical interpretation of populations would seem to fit into the soft view, since it is to an extent removed from ontological reality given that its basic units are not natural, physical entities, but units of data. Confusion and reification can result, however, when the transformative steps are ignored or forgotten.

Perhaps what unifies statistical accounts of populations are causal links, such that one may be able to, if informed well enough, trace lines of influence within the group. Maybe these lines, composed of some kind of causal "glue," are not directly perceivable by most humans, and only detectable through the power rendered by use of statistical methods. The difficulty of detection, as well as the potential for arbitrariness of categorization, however, would render such a population challenging, at the very least, to discern for on-the-ground conservation purposes. After the movement from experience to data to interpretation is completed, the question of how to know and relate to what the data correspond to looms large. On such an interpretation, the softness of the population borders on intangibility, and even incoherence.

The brief discussion of the statistical approach, despite its difficulties, brings up some relevant themes for the soft view. Among these themes is the gaining of power to obtain knowledge about hidden past phenomena that a "lumping" approach provides over a "splitting" approach. It is in some ways more useful, and likely easier to, for example, obtain a death rate for a region and include a few broad categories of harm than to attempt to identify all the precise contributory causes of death for each individual in that region. In addition to increasing the ease with which data may be gathered, a lumping or pooling approach increases the facility with which the putative entity or phenomenon of interest can be thought by any single mind. This facilitation is largely, but not entirely, epistemic in nature. There in fact appear to be at least two senses of power of special relevance to the statistical understanding of the soft view: Power for greater knowledge or explanation, and power for enhanced protection, control, or manipulation. This distinction can mark the difference between pure and applied research, theory, and practice, and other academic tensions.

An important and often overlooked point for us is that, as power to know the past is often much greater than power to know the future, attempts to conserve the past, in a literal sense, are often exercises in futility. As Gould (1990) suggests, evolutionary reconstructions in the forms of phylogenies and the lineages they represent, while interesting and informative of possible and plausible trajectories of future evolution, say little of certainty about what will occur even in the absence of human disturbance. While such methods may help us conceptualize past events and situations, their applicability to the future (and accuracy in constructing the past) is partial. Perhaps we can know something about such historical populations in a very limited sense, but it is likely that this knowledge is temporally bounded to a frame that is always no longer of conservation concern.

Such basic uncertainty about temporal status appears to be a characteristic of all conservation, and an additional issue leading to the inevitable softening of the population concept. Accurate projection of statuses from time t to $t+1$ is difficult, to say the least, and this difficulty is heightened by the ever-growing complexity of social and technological interactions with the natural environment. With a lack of ample knowledge comes lack of ability to protect what we wish to protect, at the level we wish to protect it. If it is known what has happened in the past with a certain demographic, it is often more optimistically surmised, we can at least use that knowledge to try to affect what happens in the future within a range of confidence. This assumption seems valid in the limited sense it is presented, but points to an unsettling question: Does any semblance of a population which we have significant epistemic access to only exist in the past? Can populations only exist, for us, if we rewind, replay, and pause the tape, as it were (Gould 1990)?

While it is far from clear how, for example, Minnesota conservationists and natural resource managers are conceiving of the local wolf population, the launching of a hunting and trapping season shows that it is being treated as soft. What is being referred to as softness and hardness can have many distinct meanings, but for the current discussion, to say that the wolf population is being treated as soft is to say that it is not being treated as a strongly unified entity. This treatment agrees with my view that a population of animals is not, properly speaking, a thing. However, while my argument points out the lack of enduring physical bonds within the population, I have not denied that there may likely be another kind of bonding at work within the population, what are often referred to as social bonds. While these bonds of relationship and meaning may not be strong enough to warrant the granting of the status of thing to the population, the acknowledgment of such bonds may help in making a case against the separation (lethal or nonlethal) of individuals from the population (see Bekoff 2007).

This notion of the softer population is nearing a different ecological concept—that of the community, where a "unity" composed of voluntary or semi-voluntary associations takes the place of literal, physical unity. In the case of wolves, while a common view is that it is generally beneficial for individuals to form or join packs and to take on roles within these groups, there are many documented cases of "lone wolves" who can function as wholes for long periods of time outside of such group organization. While the self-sufficient functioning of these less dependent animals is thought to be suboptimal, such cases indicate that the forming of packs is to some degree voluntary. The soft view of what we are calling populations, then—at least as it pertains to mammals and other vertebrates—is beginning to appear more closely akin to a view of the human community.

While the ecological community concept is well established (Elton 1946; Simberloff and Dayan 1991), the notion I am referring to here is more basic. It is faithful to the original meaning of the word, in the sense of values and resources shared by a group of individuals. Although the cohesiveness of plant communities, for example, cannot be discounted, this structure is not of primary concern in this work. Group dynamics are only mentioned insofar as they contribute to the well-being of sentient individuals.

On the soft view, while populations of complex organisms may not be coherently seen as physically unified things, neither can the importance of social relationships within the group be discounted, nor of the potential of political initiatives for their future protection. Such relationships, moreover, appear to amount to more than just automatic roles for many vertebrates, and imply some degree of freedom of choice. This soft view becomes, then, a double blow for any attempt to defend the lethal or nonlethal disruptive treatment of a putative population, since the lives of individuals matter not only to themselves but to others in their community. While the most important thing for members of communities is that no harm befalls them as individuals, there is added value given to each life in community from other lives within the community. This flies in the face of a more mechanistic view of populations as being able to function well despite missing many previously possessed "parts."

Although the clarity of the distinction between hard and soft populations is at this point lacking, and will hopefully be improved in the future, some progress has been made toward uncovering general differences. While the hard conception is metaphysical, the soft conception is to some degree statistical, mental, and political; in a word, pragmatic. To admit, per the soft view, that populations are perhaps not as unified as ecologists would like, would be an improvement—as would the acceptance that many presumed populations are in fact not treated as units. The pragmatism of the soft view, however, has not yet been sufficiently tempered by ethical considerations. We are still at the early stage of wanting unity while wanting to be able to destroy parts of the unit at our discretion; we want to have the freedom to lump and split at will and with violence, with little or no regard for what our senses tell us about the integrity of nature's furnishings. An understanding of diversity within single-species vertebrate groups as indicating the value of organisms has yet to be fully realized. As we perceive the path of evolution and note the differing complexities of organisms, arriving at sentience, we should treat the holders of these capacities as their expression, or potential for expression, merits.

References

Ariew, A. 2008. Population Thinking. In *The Oxford Handbook of Philosophy of Biology*, ed. M. Ruse. Oxford: Oxford University Press.

Bekoff, M. 2007. *The Emotional Lives of Animals: A Leading Scientist Explores Animal Joy, Sorrow, and Empathy—And Why They Matter*. Novato: New World Library.

Elton, C. 1946. Competition and the Structure of Ecological Communities. *The Journal of Animal Ecology* 15 (1): 54–68.

Gould, S. J. 1990. *Wonderful Life: The Burgess Shale and the Nature of History*. New York: W. W. Norton & Company.

Simberloff, D., and T. Dayan. 1991. The Guild Concept and the Structure of Ecological Communities. *Annual Review of Ecology and Systematics* 22:115–143.

CHAPTER 5

Killing and Conservation: A Case Study of the Gray Wolf

Abstract It might be claimed that such a bio-ontological analysis as the foregoing is academic, and of little import for real conservation practice. After all, conservation biology is an applied field, and gains nothing if good conservation is not advanced by theory. How, then, would a scaled-down conservation look different from what is now practiced? One way of marking the differences is by way of analysis of actions and practices that are currently taking place. We will call this the negative critique. Are there any activities currently being performed under the name of conservation that would be brought into question if assessed from an organismic perspective? First, we will examine activities that would likely be evaluated differently if such an ecological downscaling was adopted. Then we will present a vision of what a more positive organismic conservation might include.

Keywords Gray wolf • Culling • Habitat • Attitude • Hunting

It might be claimed that such a bio-ontological analysis as the foregoing is academic, and of little import for real conservation practice. After all, conservation biology is an applied field, and gains nothing if good conservation is not advanced by theory. How, then, would a scaled-down

L. B. DeVaney, *Animal Conservation Ethics and the Population Problem*, The Palgrave Macmillan Animal Ethics Series, https://doi.org/10.1007/978-3-031-91891-9_5

conservation look different from what is now practiced? One way of marking the differences is by way of analysis of actions and practices that are currently taking place. We will call this the negative critique. Are there any activities currently being performed under the name of conservation that would be brought into question if assessed from an organismic perspective? First, we will examine activities that would likely be evaluated differently if such an ecological downscaling was adopted. Then we will present a vision of what a more positive organismic conservation might include.

From a mainstream conservation standpoint, prime targets for cessation (or at least mitigation) would include deforestation, urban sprawl, and pollution, among others. An added component to organism-centric conservation, and one which is much more easily defined than any of the above practices, is killing of animals by hunting and trapping. Such a reconsideration would go beyond Diamond's universally accepted "evil quartet" component of "overkill" (1989), to ceasing all unnecessary killing. Given an organism-centric approach to conservation, what counts as necessary and unnecessary killing would likely come into clearer resolution. Within the current schema, killing practices can be placed in three categories in terms of legal and governmental support: Illegal, legal-voluntary, and legal-mandated. The first type is more generally known as poaching, the second as hunting and trapping, and the third as culling. Although poaching is strongly disliked and punishable by fines and jail time, it can be transformed into something allowed and encouraged by governmental authorities. As an example, consider the following case study centering on the federal delisting of gray wolves and the Minnesota Department of Natural Resources (DNR) mandate to instate a wolf hunt.

A Final Rule to remove Endangered Species Act (ESA) protection from the Western Great Lakes Distinct Population Segment (DPS) of wolves was passed by the United States federal government on December 28, 2011. The Western Great Lakes DPS is defined as gray wolves inhabiting the states of Minnesota, Wisconsin, and Michigan. The Final Rule for ESA removal took effect on January 27, 2012, 30 days after its announcement in the *Federal Register* (USDI 2011). Viewed at the species level, gray wolves are a great conservation success story and provide a model for questioning how we ought to engage in conservation after a measure of success has been achieved. With increases in wolf numbers has come an expansion of wolf territories into areas more heavily populated by humans, which creates potential for human–wolf conflict. Although the wolf has gained esteem in recent decades as a symbol of wilderness, wolves are beginning to challenge assumptions about their favored territorial areas and food sources, and because of this are increasingly attracting negative attention from both citizens and governmental agencies. For context, I will briefly discuss some relevant historical, biological, and ecological aspects of wolves and their environments below.

Although wolf bounties have existed in North America since the early seventeenth century, wolf numbers remained stable during the next few centuries due to low interest from fur traders and low settlement rates (Mech 1970; Bruskotter et al. 2007). Numbers began to decline, however, as settlement, forest logging, and land clearing rates increased in the late nineteenth century (Stearns 1990; Mladenoff et al. 1995). While being adversely affected by these impacts on habitat and prey, wolves were killed directly and intentionally by way of firearms, poisons, and traps because of their perceived threat to livestock and humans (Young and Goldman 1944; Mech 1995). Numbers are thought to have been lowest in the late 1950s, by which time, save for remnant packs in northern Minnesota and Lake Superior's Isle Royale, wolves were all but absent from the lower 48 states (Mech 1995). The onset of federally supported control in 1915 is thought to have contributed greatly to the decline over the following decades (Di Silvestro 1985; Bruskotter et al. 2007).

During the 1960s, public attitudes toward wolves began to change with the help of the broader environmental movement. This change in attitudes was augmented by new research techniques and legislation. Radio tracking, for example, an important technique for studying animals with large ranges and low numbers, was developed in the early 1960s (Cochran and Lord 1963; Mech 1995), and in 1966, Congress passed the Endangered Species Preservation Act; wolves were listed the following year (Bruskotter et al. 2007). This act, however, did little to protect wolves in Minnesota, as both an open season and Directed Predator Control Program were observed from 1965 to 1974. Around 250 wolves were killed each year during this time span (United States Fish and Wildlife Service, USFWS 2011).

The more familiar, and substantial, ESA was passed in 1973, and recovery teams were appointed by the United States Fish and Wildlife Service (USFWS) to three different gray wolf subspecies, including Minnesota's eastern timber wolf or buffalo wolf (Mech 1995). Although illegal killing persisted over the next few years, wolves soon began to benefit from ESA listing, with Minnesota packs beginning to recolonize parts of both Wisconsin and Michigan's Upper Peninsula. The wolves of these three states would come to be known as the Western Great Lakes DPS, although some Minnesota wolves have since spread into the Dakotas as well (Licht and Fritts 1994).

During the early recovery years of the 1970s and 1980s, the idea of wolves as being dependent on wilderness was further solidified through studies linking low road density on the landscape to high wolf population density. Thiel, for example, found that wolves had only recolonized parts of Wisconsin with less than 0.6 km/km^2 road density (1985). This figure, however, is no longer indicative of the areas populated by wolves in the Western Great Lakes region. Although during the 1970s wolves were found primarily in the northeastern corner of Minnesota, they now occupy most of the state, as well as contiguous states to the east and west. As early as 1994, breeding packs were observed within 90 km of Minneapolis/St Paul, and individual wolves were seen within 30 km of this major metropolitan area (Wydeven 1994; Mech 1995). Along with this large increase in range has come an accompanying increase in numbers, from an estimated 500 to 1000 in 1973, to nearly 3000 in 2008 (USFWS 2011). The estimate in 2020 was 2696 wolves (Erb and Humpal 2020: 7).

While it may be true that wolves prefer areas with small numbers of humans and low anthropogenic impact, their adaptability has been proven in the decades since the 1960s. This, overall, is good news; but it has the potential to inspire overconfidence about the persistence of wolves through time. The danger of excessive optimism is underscored by the recent ruling to remove ESA protection from the Western Great Lakes DPS. The USFWS, however, assured "survival of the gray wolf population in Minnesota and a population of 100 or more wolves in Wisconsin/Michigan for a minimum of five consecutive years" (USFWS 2011: 1).

Wolves are known to be habitat generalists; worldwide, they are found in prairie, forests, mountains, wetlands, and farmlands. In the Western Great Lakes region, they are becoming increasingly common in mixed forest-agriculture areas and are expanding their range toward suburban and urban communities. However, it is believed that full expansion into densely human-populated areas will be severely limited by human–wolf conflict and urban threats such as automobile traffic (USFWS 2006). Territories can span 80 or more square kilometers, and wolves have been known to travel 48 kilometers a day in search of white-tailed deer (*Odocoileus virginianus*) or other favored prey such as moose (*Alces alces*),

North American beaver (*Castor canadensis*), or snowshoe hare (*Lepus americanus*). When seeking larger prey, wolves tend to hunt the young and old to minimize energy use and risk of injury.

The wolf's family group is the pack, usually led by an alpha pair. Normal pack size is four to eight in the Midwest but can be as high as 30 or more in Canada and Alaska. Litters often consist of five or six pups, produced once yearly by the alpha pair (Mech 1970). The breeding season is late January through March, with gestation around two months. The mother often digs a den into soil, as deep as three meters, but may instead choose a cave, hollow log, or vacant beaver lodge. Wolves are weaned at about six weeks in the summer. Soon after weaning, the pups are moved to a rendezvous/nursery site, which they leave in the autumn. Adult size is reached by one year, and wolves can weigh between 27 and 55 kilograms. Wolves can reach 13 years of age in the wild, and 15 or more in captivity (USFWS 2006).

As the above data indicate, habitat requirements for wolves are not as specific as was once observed. Wolves are no longer widely held to be wilderness specialists and have shown adaptability by expanding their ranges into agricultural and other human-impacted areas. However, it is crucial that this potential habitat is not only adaptable to wolves themselves but can also sustain enough natural prey. If it cannot, then this raises the likelihood that wolves will either actively seek out or happen upon domesticated animals. Because the mixed forest-agricultural landscape common in much of the midwestern United States often includes a tempting combination of tree cover, open pastureland, and livestock, depredation on domestic animals, including pets, sometimes occurs in these areas.

There are conflicting beliefs as to whether, all else being equal, wolves prefer wild or domesticated prey, and this conflict may be related to nongeneralizable differences in the preferences of individual wolves. Some studies, however, support wolf predation on livestock as linked to lack of wild prey (Mech et al. 1988; Meriggi and Lovari 1996; Treves et al. 2003). Whatever the case, it is agreed that some wolves prey on livestock—often outside of known territories and ranges (Treves et al. 2003)—resulting in heavy costs both to farmers and wolves. Payment schemes exist to compensate for damages incurred; however, such payments can divert funds from other conservation-related activities and can function as a reason for instating revenue-generating public hunting seasons in their place (Thompson 1997; Treves et al. 2009).

Concerning the future of wolves in Minnesota, what monitoring efforts will consist of, what the metrics for success will be, and the level of federal involvement after the five years are up remain to be seen. Although it seems that the relevant state and tribal agencies are welcoming this increased accountability for wolf management with open arms, a significant amount of uncertainty exists concerning future laws, policies, and enforcement at the state and local levels. What is certain at this point, however, is that the first DNR-sanctioned wolf-hunting season in Minnesota history started on November 3, 2012. This was despite roughly 80% opposition from the 7351 responses received to a recent DNR-administered survey. In addition, those on both sides of the issue have pointed out what they perceive to be methodological errors, including lack of state specificity of respondents and the addressing of a large number of questions to prospective hunters rather than the general interested public (Smith 2012b).

In spite of this show of significant public disapproval for the wolf hunt, given by the unambiguous survey question "Do you support hunting and trapping for wolves in Minnesota?", the DNR website conflictingly states that "The survey was not a referendum on whether to hold the season but to elicit comments on how the season would be implemented" (MNDNR Wolf Management). Wildlife chief Dennis Simon voiced the unyielding, settled-in-advance stance of the DNR when he said, "The Legislature and governor directed us to have a wolf season. So we will have a season" (Smith 2012b). The "target harvest" was set at 400 wolves, while those who might worry about numbers were informed that the "minimum goal" for the population is 1600. If numbers fell below this threshold, "appropriate management actions" would be taken "to reverse the decline" (Ibid.).

Survey responses were not the only early indicators of public disapproval, however. On August 9, 2012, a petition was filed by the group Howling for Wolves requesting that the planned hunting and trapping season be cancelled. The DNR denied this petition on October 2, 2012. A lawsuit was also filed with the Minnesota Court of Appeals and Minnesota Supreme Court by Howling for Wolves and the Center for Biological Diversity against the Minnesota DNR and Commissioner Tom Landwehr. In both cases, the injunction was denied, and the season unfolded as initially planned (Ibid.).

Although this season[1] marked the first time in Minnesota history that a wolf hunting and trapping season was instated, in addition to the alarming fact that 6000 licenses were offered—two times the estimated number of the then-current Minnesota wolf population—International Wolf Center founder and world-renowned wolf expert L. David Mech has boldly stated, "That isn't going to hurt the population" (Smith 2012a). This statement points to the absurdity of the idea of a population experiencing pain. If removal of roughly 13% of the population's "body" would not cause pain, it would seem that nothing would. The question of why a leading expert would use such imprecise language in a context this serious looms large.

Mech has even deemed the plan "conservative" in a recent legislative hearing, banking on the challenge of "successfully" hunting wolves to keep the kill numbers low (MNDNR Wolf Management). This statement, however, fails to account for data that Mech is likely aware of, given his 50-plus years of wolf research. First of these pieces of data is the general high kill rates of trappers. According to figures published by the Minnesota DNR on trapping outcomes from 1990 to 2005, the mean estimated kill number per trapper per year of three species of canids (coyote [*Canis latrans*], red fox [*Vulpes vulpes*], and gray fox [*Urocyon cinereoargenteus*]) was five individuals (Dexter 2005: 245). Although it may be assumed that rates of wolf trapping "success" will be lower than with other canids, the elusiveness of wolves sensing human danger that is expected for hunting is not a reasonable assumption for trapping.

Second is a point directly spoken to by Mech and discussed above: The fact that the gray wolf is no longer solely an animal of wilderness. Livestock predation by wolves and unease of some humans with wolf proximity are precisely the problems at issue, according to many. The likelihood, however, that increasing adaptation to human-altered landscapes will make many wolves easier to kill than the traditionally conceived "wilderness wolf" is seemingly overlooked. Knowledge of this accessibility may itself draw out many more hunters than would be tempted by travel through vast expanses of forest that many seem to be envisioning.

[1] The season was divided into early and late halves. "Early" was hunting-only and lasted from November 3 to 18, 2012. "Late" allowed both hunting and trapping and lasted from November 24, 2012 to January 31, 2013.

By 2021, the gray wolf had been delisted from everywhere in the lower-48 United States, excluding the Mexican wolf subspecies. In 2022, however, federal protection was reinstated for all gray wolves outside of the Northern Rocky Mountain Region (USFWS 2023). This included the classification of wolves in Minnesota as threatened effective February 10, 2022 (MNDNR Wolf Management 2024). As of 2024, pressure is mounting yet again to reverse this protection and have another hunt.

References

Bruskotter, J. T., R. H. Schmidt, and T. L. Teel. 2007. Are Attitudes Toward Wolves Changing? A Case Study in Utah. *Biological Conservation* 139:211–218.

Cochran, W. W., and R. D. Lord Jr. 1963. A Radio-Tracking System for Wild Animals. *Journal of Wildlife Management* 27:9–24.

Dexter, M. 2005. *Trapping Harvest Statistics.* Saint Paul: Minnesota Department of Natural Resources, Division of Fish and Wildlife.

Di Silvestro, R. L. 1985. The Federal Animal Damage Control Program. In *Audubon Wildlife Report*, ed. R. L. Di Silvestro, 130–158. New York: National Audubon Society.

Diamond, J. 1989. Overview of Recent Extinctions. In *Conservation for the Twenty-First Century*, ed. D. Western and M. Pearl. New York: Oxford University Press.

Erb, J., and C. Humpal. 2020. *Minnesota Wolf Population Update 2020.* St. Paul: MN DNR.

Licht, D. S., and S. H. Fritts. 1994. Gray Wolf (*Canis lupus*) Occurrences in the Dakotas. *American Midland Naturalist* 132 (1): 74–81.

Mech, L. D. 1970. *The Wolf: The Ecology and Behavior of an Endangered Species.* Minneapolis: University of Minnesota Press.

Mech, L. D. 1995. The Challenge and Opportunity of Recovering Wolf Populations. *Conservation Biology* 9 (2): 270–278.

Mech, L. D., S. H. Fritts, and W. J. Paul. 1988. Relationship between Winter Severity and Wolf Depredation on Domestic Animals in Minnesota. *Wildlife Society Bulletin* 16:269–272.

Meriggi, A., and S. Lovari. 1996. A Review of Wolf Predation in Southern Europe: Does the Wolf Prefer Wild Prey to Livestock? *Journal of Applied Ecology* 33:1561–1571.

Minnesota Department of Natural Resources. 2024. Wolf Management | Minnesota DNR. https://www.dnr.state.mn.us/wolves/index.html. Accessed 9 December 2024.

Mladenoff, D. J., T. A. Sickley, R. G. Haight, and A. P. Wydeven. 1995. A Regional Landscape Analysis and Prediction of Favorable Gray Wolf Habitat in the Northern Great Lakes Region. *Conservation Biology* 9 (2): 279–294.

Smith, D. 2012a. It's Open Season on Rules for Minnesota Wolf Hunt. *Star Tribune*, May 22, 2012.

Smith, D. 2012b. Minnesota DNR Survey: Most Oppose Wolf Hunt. *Star Tribune*, June 22, 2012.

Stearns, F. 1990. Forest History and Management in the Northern Midwest. In *Management of Dynamic Ecosystems*, ed. J. M. Sweeney. West Lafayette: The Wildlife Society, North Central Section.

Thiel, R. P. 1985. Relationship between Road Densities and Wolf Habitat Suitability in Wisconsin. *American Midland Naturalist* 113:404–407.

Thompson, B. H., Jr. 1997. The Endangered Species Act: A Case Study in Takings and Incentives. *Stanford Law Review* 49:305–380.

Treves, A., L. Naughton-Treves, E. K. Harper, D. J. Mladenoff, R. A. Rose, T. A. Sickley, and A. P. Wydeven. 2003. Predicting Human-Carnivore Conflict: A Spatial Model Derived from 25 Years of Data on Wolf Predation on Livestock. *Conservation Biology* 18 (1): 114–125.

Treves, A., R. L. Jurewicz, L. Naughton-Treves, and D. S. Wilcove. 2009. The Price of Tolerance: Wolf Damage Payments after Recovery. *Biodiversity Conservation* 18:4003–4021.

U.S. Department of the Interior. 2011. Endangered and Threatened Wildlife and Plants; Revising the Listing of the Gray Wolf (*Canis lupus*) in the Western Great Lakes. *Federal Register* 76(249):81666–81726. https://www.federalregister.gov/documents/2011/12/28/2011-32825/endangered-and-threatened-wildlife-and-plants-revising-the-listing-of-the-gray-wolf-canis-lupus-in. Accessed 8 December 2024.

U.S. Fish and Wildlife Service. 2006. *Gray Wolf Biology Questions and Answers.* Washington, DC: USFWS. https://www.fws.gov/sites/default/files/documents/2024-04/1.pdf. Accessed 8 December 2024.

U.S. Fish and Wildlife Service. 2011. *Gray Wolf Recovery in Minnesota, Wisconsin, and Michigan.* Washington, DC: USFWS. https://www.wisconsinwolffacts.com/forms/fedwolfplan.pdf. Accessed 8 December 2024.

U.S. Fish and Wildlife Service. 2023. *USFWS Gray Wolf Statement.* Washington, DC: USFWS. https://www.fws.gov/sites/default/files/documents/2023%20USFWS%20Gray%20Wolf%20Statement.pdf. Accessed 8 December 2024.

Wydeven, A. P. 1994. Travels of a Midwestern Disperser. *International Wolf* 4 (1): 20–22.

Young, S. P., and E. A. Goldman, eds. 1944. *The Wolves of North America.* New York: Dover Publishers.

PART II

Ethics and Conservation

The second part of this book is an attempt at beginning to bridge the gap between theory and practice. Where the first part remained focused on the meaning and metaphysical issues surrounding conservation biology, this second part is oriented toward addressing ethical issues associated with human treatment of wild animals. Description of wolf biology and the historic Minnesota wolf hunting season of 2012–2013 in Chap. 5 moves into an analysis of rationales for killing. Three main lines of argument attempting to justify killing are reviewed and their weaknesses exposed.

I then progress into an attempt at balancing welfare and ecology. This is performed through scrutiny of a variety of bases for ascription of value. A thought experiment is offered for helping think through possible scenarios in which a creature in need of aid may be discovered and potential responses. This provides a transition into part three on practice.

CHAPTER 6

Questioning Killing

Abstract Aside from intentional, direct killing from humans, wolves, like many mammals, suffer from unintentional, direct killing and injury, habitat loss, intraspecific competition, disease, and starvation from lack of prey (Mech 1970, 1977; Keith 1983; Fuller 1989; Mladenoff et al. 1995). Effects on the landscape and in the surrounding environment, and therefore to potential habitat, can include logging; barriers and obstacles such as roads, railways, trails, and fences; mining pits and quarries; climate change; and fires (Vors et al. 2007; Wilcove and Wikelski 2008). This is a daunting list of candidates for treatment, especially considering the limited budgets conservation organizations operate on coupled with the uncertainty of success. Although each type of anthropogenic activity listed above poses a threat to wolves and other species, underscoring the importance of laws and policies addressing the entire suite of risks, direct intentional killing without cause is viewed by the author as a particularly vicious activity for humans to engage in, and one that is antithetical to the most elementary of concerns for environmental protection and care. The magnitude of this viciousness is compounded, moreover, through the offering and acceptance of an opportunity to hunt members of a sentient species less than a year after having been essentially declared "safe."

L. B. DeVaney, *Animal Conservation Ethics and the Population Problem*, The Palgrave Macmillan Animal Ethics Series, https://doi.org/10.1007/978-3-031-91891-9_6

Keywords Killing • Native species • Invasive • Extinction • Stochasticity

Aside from intentional, direct killing from humans, wolves, like many mammals, suffer from unintentional, direct killing and injury, habitat loss, intraspecific competition, disease, and starvation from lack of prey (Mech 1970, 1977; Keith 1983; Fuller 1989; Mladenoff et al. 1995). Effects on the landscape and in the surrounding environment, and therefore to potential habitat, can include logging; barriers and obstacles such as roads, railways, trails, and fences; mining pits and quarries; climate change; and fires (Vors et al. 2007; Wilcove and Wikelski 2008). This is a daunting list of candidates for treatment, especially considering the limited budgets conservation organizations operate on coupled with the uncertainty of success. Although each type of anthropogenic activity listed above poses a threat to wolves and other species, underscoring the importance of laws and policies addressing the entire suite of risks, direct intentional killing without cause is viewed by the author as a particularly vicious activity for humans to engage in, and one that is antithetical to the most elementary of concerns for environmental protection and care. The magnitude of this viciousness is compounded, moreover, through the offering and acceptance of an opportunity to hunt members of a sentient species less than a year after having been essentially declared "safe."

As has been noted, killing is often justified in terms of claimed benefits to aggregates or groups of organisms. This type of argument is most often framed as a scientifically based argument, whereas arguments that appeal to the welfare of individuals are frequently labeled as ethical or moral, nonscientific, and/or sentimental. The following are some common lines of reasoning that are used in support of arguments for killing aiding groups of animals, given in three statements:

1. If directed at members of a nonnative species, culling will reduce potentially invasive numbers, also reducing negative predatory or competitive impacts on native species.
2. If directed at members of a native herbivorous species, culling will reduce pressure on limited resources used by the prey species, thereby doing good for the target population(s) and/or species in the long run.

3. If directed at members of a native omnivorous or carnivorous species, culling will reduce pressure on limited resources used by the predator species, thereby doing good for the target population(s) and/or species in the long run.

These are, of course, limited arguments in the sense that I am only providing the basic form of justifications that can take varying levels of complexity and specificity. However, since they are effectively templates for many similar rationales that have been given by natural resource agencies and conservation organizations in the recent past, they warrant further scrutiny. Although a general evaluation of the validity of the given rationales cannot perfectly align with the nuances of each specific case to which they might apply, perhaps it can lead us to an enhanced understanding of their overall fit. Let us, then, take a closer look at these lines of thought and their main assumptions, imagining them as already solidified positions and working backwards to investigate potential supporting rationales and alternative perspectives.

The first statement is perhaps the least controversial of the three and assumes that possessing the status of temporal precedence is preferable to appearing later on the scene, however this lateness is caused. Additionally, rather than taking the routes of either not interfering or interfering nonlethally, it has been deemed better to kill. We can see here, from a very brief assessment, that not only facts but also value judgments are in play concerning native versus nonnative designations. Although judging the relative merit of available options may not have been raised to the level of explicitness, implicit choices have at least been made. The judgments leading to such choices are not necessarily ethical in character—for example, they may have been derived from economic or aesthetic considerations—but they clearly reflect the presence of values and preferences as motivating factors. This is not to claim that the basis for such decisional influences is arbitrary or not influenced by external evidence, but that internal psychological processes play a significant role in their structure.

What are some possibilities that could influence the arrival at such attitudes in favor of killing nonnative species? One is that the new species is or will become invasive, gaining an ecological foothold that could negatively affect natives at some level for a duration of time. In a case of native–nonnative conflict, however, the competition provided by the current or possible future invasive species could improve the overall fitness of native species in the long run, thereby providing an increased chance of evolutionary success in a changing environment. In addition, it is possible that a clearer mutualism between the species could develop and emerge over time, such that many nonnative species come to be regarded as ecologically valuable (Schlaepfer et al. 2011).

There is also an apparent, inherent risk that a nonnative species will cause certain native species to decline severely, or even contribute to extinction events. Since it is unclear when and whether an invading species will have success at the expense of a preferred native, the argument goes, it is better to err on the side of caution and assume that it will have such a harmful effect. A proactive response to such an assumption, it would seem, would be to quickly destroy as many members of the offending species as possible. Even if such a feared harmful effect is experienced, however, in most cases it is more likely that the effect will be small to moderate than that the problem species will so efficiently and harmfully exploit a new potential niche against a long-established native that the latter is driven to extinction.

Some questions, at this point, seem to be: (1) What is the extent of the relevant risk? (2) Can anything be done to lessen it? and (3) If the risk is deemed great and something can be done to lessen it, is killing justified as a remedial measure? In addition, and in hindsight, if the native species is in fact driven to hardship by the nonnative, on an evolutionary reading this would seem to indicate a subpar fit within the niche the former was thought to occupy. The newfound success of the nonnative species, in any case except that of overt mutualism, is very likely to count against it because of the relative values and disvalues that tend to be placed on native and nonnative species. Notice, moreover, that members of the nonnative species as conceived in the foregoing line of reasoning, taken on their own terms and given their location, are apparently not thought of as having a good; no appeal is made to goods or interests of either the population or individual members thereof.

In the second statement, the species being targeted is specified as both native and herbivorous, and it is assumed that the population has reached or will soon reach an unsustainable level. Generally, this is thought to be the case because it is believed that resources are being used at a rate which the local environment cannot support—in ecological terms, carrying capacity has been or will soon be reached. An additional rationale commonly attached to such a case, often implicit, is that the supposedly overabundant species is a favored game animal, and/or that it is the hunter's responsibility to take the place of declining natural predators for keeping the population in check.

On the face of it, much of the above reasoning seems sound. A major problem with such an argument, however, is lack of account of uncertainty and randomness. This presence of unpredictability is often referred to in ecology theory as stochasticity. Although attempts are continually made to incorporate stochasticity (as opposed to determinism) into models, there is doubt concerning the extent to which this phenomenon of randomness can be modeled. An important philosophical question centers on the accuracy of modeling events that befall concrete, nonrandom individual lives as though they occur to abstracted, random individuals composing a modeled population. Although strides have undoubtedly been made in predictive ecology, in most cases we have insufficient power to know in advance the fates of real individuals who are dependent upon a certain resource. Moreover, models of resource dependence often do not support the possibility of switching to alternative resource types in times of hardship.

Rather than the entire population of note going extinct from starvation, for example, it is much more likely that some individual organisms will die. If the Darwinian theory of natural selection (Darwin 1859) is roughly correct, then mortality rates are measures of relative fitness, such that when individuals die of natural causes there are fitness-related reasons for their deaths. Therefore, unless the population is a true unit, there will be differential rates of birth and death according to fitness. During tough times, more of the unfit will die than the fit.

Assuming natural causes begs the question of the extent to which the environment, and the causes of death produced therein, is natural. If it is found that there is a significant unnatural—for example, technological—component, this complicates things by bringing in a host of factors that wildlife have not had sufficient opportunity to evolve with and become fit in relation to. This discussion broadens out to a consideration of the Anthropocene as being essentially destructive (Moellendorf 2020). Without going to this extreme, however, we may consider the ways in which technology has directly affected and is directly affecting the natural world. One of the most obvious ways that this has been and is happening is through hunting.

Hunting with firearms and other technologies has fundamentally altered the ways in which humans interact with animals, and the biotic effects these interactions have. Although rare cases of mercy killing of a starving, injured, or ill animal by hunters do exist, hunters aiming to procure trophies and/or meat usually select for the most fit individual—which amounts to a choice against fitness for the organism and group. If we add to this the skewing of sex ratios that often follows from biased killing of males, we no longer have such a strong argument for hunting plant-eaters (Cohn 1999). It seems there is now a very different picture wherein the fittest members of a species are being killed, often cruelly, to the detriment of both genetic diversity and fitness.

Many herbivores, however, do not "enjoy" the status of game animals, but are instead adversely portrayed as vermin. This value-laden term can be applied for many reasons, including judgments of disease, crop or property damage, or other danger or harm, but there seems little room for positive interpretation once this label is affixed to a group of creatures. If something is vermin (or a pest, a scavenger, a nuisance, exotic, or invasive) (Nagy and Johnson 2013: 1) it is totally undesirable, and therefore killing of such can be less selective than in sport hunting, wholesale, and carried out by individuals, businesses, and government agencies. Organisms believed to belong in this category often include small mammals—especially rodents—but can also include larger, predatory animals found more aesthetically appealing, such as coyotes. Given that there is little value of any kind attached to vermin, limitations to their destruction are often entirely lacking, with no citing of population-level benefits given as justification. After all, would it not seem strange to talk of a healthy mouse population? But if health is a quality experienced by populations of larger mammals, why should it not be had by, and important to, mice, such that we should be looking out for their interests, genetic integrity, etc.? If shrubs, why not shrews?

The problems do not stop here, however, if we attempt to use the third statement given above in support of killing predators. The general points made about lack of certainty hold in this scenario as well, and are likely intensified due to the increased complexity of an additional trophic level being factored in. In this case as in the above cases, evidence that could lead to precise knowledge, in advance, of the effects on either particular individuals or entire populations is lacking, as is knowledge of the outcomes culling efforts would achieve. This is not a problem, moreover, that waiting for more data to project into the future will solve. Ought we, then, to commence mass killing efforts despite this uncertainty, in faith that good effects will outweigh the massive amount of harm imposed?

There is reason for thinking we ought not. It is one thing, for example, to target individual wolves or coyotes that are believed to have preyed on livestock, but something else entirely to randomly destroy large numbers of individuals, who may or may not be blameworthy, in the hope that a desired goal will be met through our actions. The feeding habits of higher predators are known to vary; examples of this variance include individual man-eating tigers (*Panthera tigris*) and lions (*Panthera leo*) that have killed hundreds of people, compared with most members of the species that will never harm a human. The possibility exists, on the one hand, that large-scale killing of specialist predators will not only release pressure on favored prey but cause a rapid increase in numbers. While Cohn aptly points out that assuming lightened predator pressure alone would cause a boom in prey numbers amounts to a "gross oversimplification" (1999: 120), there is little doubt that predators have historically played a major role in controlling prey numbers. Indeed, in the case of human predators, sometimes this control has been to the demise of entire species.

On the other hand, it is sometimes the case that culling of predators appears to cause a marked *decrease* in prey numbers. For example, Orford cites a Kruger Park, South Africa, management plan that led to the killing of 335 lions and 297 hyenas in a three-year period (Ibid. 160). It was believed that such drastic action would cause the flagging numbers of zebra (genus *Equus*) and wildebeest in the park to increase, but instead they declined. Absurdly, prior to the predator culling effort, roughly 4000 wildebeest were culled because of fear of overgrazing. After all this bloodshed, management decided the herbivores were in a phase of population decline and they were then left alone (Ibid.). The above scenario calls into serious question assumed causal linkages and the presumption of simple, predictable, controllable ecological relationships.

In recent years, reasons for killing of predators, as in the wolf example, have often been supported by appeals to protection of livestock. While wolf (and other predator) numbers, through hunting and trapping, habitat loss, and various other anthropogenic causes, may no longer be seen as sufficient to keep the deer herd "in check," they are sufficient to affect confined and unprotected herds of domestic animals. Justification of killing from animal agriculture, however, is especially absurd from an environmental perspective. The United Nations Food and Agriculture Organization (FAO) has referred to animal agriculture as among the most serious causal factors for environmental degradation (Waldau 2010). The FAO report *Livestock's Long Shadow* states that the livestock sector "is by far the single largest anthropogenic user of land"; "is … responsible for 18 percent of greenhouse gas emissions … a higher share than transport"; and "is probably the largest sectoral source of water pollution." In addition, "Livestock now account for about 20 percent of the total terrestrial animal biomass, and the 30 percent of the earth's land surface that they now pre-empt was once habitat for wildlife" (Steinfeld et al. 2006: xxi–xxiii). Note that this critique, as scathing as it appears, does not refer to any concerns for the welfare or rights of the domestic animals themselves.

Making our way back to wolves, perhaps the most damning statements concerning the benefit of killing predators come from Mech himself. In reference to the often cited "trophic cascade effect," a top-down model of ecological control believed to stem from the influence of large predators on prey numbers and behavior and progressing throughout the ecosystem, Mech states that "we do not claim to know whether the wolf's effects are positive or negative, what its net effect is, or whether the effects are of any great consequence ecologically" (2012: 144). It would seem to follow from this that, if we do not know the effect top predators have ecologically, then we do not know what will be gained by controlling them. Regarding the attempt to make sweeping claims about predator–prey relationships and effects, Mech and Peterson wrote that "the alleged 'control' of prey populations by predators, for example, might be interpreted in at least six ways, depending on the definition used" (2003: 146). The authors go on to state that "Limited by predation, prey populations will rise and fall at irregular intervals based on demographic and environmental factors that influence losses to predators" (Ibid.: 155). These admissions of uncertainty can be seen as not only challenges to the popular conservation concept of keystone species, but the pervasive ideology of ecological organization and unity.

This brief study of wolves and their prospects in Minnesota and neighboring states, coupled with a critical assessment of killing practices, reveals the potential for a dramatic shifting of conservation categories and contexts. Perhaps the most striking shifts, in the case of the wolf, can be seen in the change from endangered status to claimed abundance and the opportunity for drastically different activities that this has opened. If the USFWS is right in its appraisal of the gray wolf, a rebound from species-level danger can happen in a short amount of time. Will such rapid transitions indicating conservation success continue to be viewed as opportunities to capitalize on a newly minted resource, or as opportunities to exercise character qualities, such as restraint and compassion, that both drive our own moral evolution and indicate respect for the lives of others? This is a critical question for conservation in the twenty-first century and beyond.

References

Cohn, P., ed. 1999. *Ethics and Wildlife*. Lewiston: The Edwin Mellen Press.

Darwin, C. 1859. *On the Origin of Species by Means of Natural Selection, or The Preservation of Favoured Races in the Struggle for Life*. London: John Murray.

Fuller, T. K. 1989. Population Dynamics of Wolves in North-Central Minnesota. *Wildlife Monograph* 105:1–41.

Keith, L. B. 1983. Population Dynamics of Wolves. In *Wolves in Canada and Alaska*, ed. L. N. Carbyn. Ottawa: Canadian Wildlife Service. Report Series 45.

Mech, L. D. 1970. *The Wolf: The Ecology and Behavior of an Endangered Species*. Minneapolis: University of Minnesota Press.

Mech, L. D. 1977. Wolf-Pack Buffer Zones as Prey Reservoirs. *Science* 198 (4314): 320–321.

Mech, L. D. 2012. Is Science in Danger of Sanctifying the Wolf? *Biological Conservation* 150:143–149.

Mech, L. D., and R. O. Peterson. 2003. Wolf-Prey Relations. In *Wolves: Behavior, Ecology, and Conservation*, ed. L. D. Mech and L. Boitani. Chicago: University of Chicago Press.

Mladenoff, D. J., T. A. Sickley, R. G. Haight, and A. P. Wydeven. 1995. A Regional Landscape Analysis and Prediction of Favorable Gray Wolf Habitat in the Northern Great Lakes Region. *Conservation Biology* 9 (2): 279–294.

Moellendorf, D. 2020. Three Interpretations of the Anthropocene: Hope and Anxiety at the End of Nature. *Ethics, Politics & Society* 3:133–153.

Nagy, K., and P. D. Johnson II, eds. 2013. *Trash Animals: How We Live with Nature's Filthy, Feral, Invasive, and Unwanted Species*. Minneapolis: University of Minnesota Press.

Schlaepfer, M. A., D. F. Sax, and J. D. Olden. 2011. The Potential Conservation Value of Non-Native Species. *Conservation Biology* 25 (3): 428–437.

Steinfeld, H., P. Gerber, T. Wassenaar, V. Castel, M. Rosales, and C. de Haan. 2006. *Livestock's Long Shadow: Environmental Issues and Options*. Rome: United Nations, Food and Agricultural Organization.

Vors, L. S., J. A. Schaefer, B. A. Pond, A. R. Rodgers, and B. R. Patterson. 2007. Woodland Caribou Extirpation and Anthropogenic Landscape Disturbance in Ontario. *Journal of Wildlife Management* 71 (4): 1249–1256.

Waldau, P. 2010. *Animal Rights: What Everyone Needs to Know*. New York: Oxford University Press.

Wilcove, D. S., and M. Wikelski. 2008. Going, Going, Gone: Is Animal Migration Disappearing? *PLoS Biology* 6 (7): e188.

CHAPTER 7

Balancing Welfare and Ecology

Abstract We have seen that significant problems exist regarding the conventional conceptualization of one of the main levels treated in conservation biology, the population. It must be concluded, then, that the ontological status of the population as a thing, although not definitively invalidated, is presently in question. This may be construed as a theoretical point of little practical importance, such that, despite apparent problems, conservation biologists will go about business as usual. However, the question of what differences in conservation practice are likely to result from skepticism about populations, should this increase, bears asking.

Keywords Animal welfare • Ethical • Moral • Conativism • Will • Restoration

We have seen that significant problems exist regarding the conventional conceptualization of one of the main levels treated in conservation biology, the population. It must be concluded, then, that the ontological status of the population as a thing, although not definitively invalidated, is presently in question. This may be construed as a theoretical point of little practical importance, such that, despite apparent problems, conservation

L. B. DeVaney, *Animal Conservation Ethics and the Population Problem*, The Palgrave Macmillan Animal Ethics Series,
https://doi.org/10.1007/978-3-031-91891-9_7

biologists will go about business as usual. However, the question of what differences in conservation practice are likely to result from skepticism about populations, should this increase, bears asking.

One potential and significant change is that the granting of rights and acceptance of responsibilities for wildlife would shift down in scale from species and population levels, with the attending effect that individual welfare would be taken more seriously by many conservationists. Note that it does not follow that there would be a concurrent lack of emphasis placed on either groups or unique kinds of organisms; in such a situation it would be possible to have both large- and small-scale conservation, as long as the large-scale type adequately took individual welfare considerations into account. If the coherence and cohesiveness of larger or more complex biotic wholes was deemed suspect, however, then many of the benefits currently reserved solely for SONEs might be extended to organisms, such that individual welfare was no longer trumped by the supposed good of the collective. The principle of "do no harm," which is currently often interpreted as only applying to the population or higher level, might then be re-interpreted as applying primarily at the organismal level, with the implied proviso that either sacrificing of lives or failure to attend to needs or protection of large numbers would be unacceptable. In other words, a possible consequence of an increased emphasis on the organismal level would be a raising of the ethical status of the individual in relation to that of the population or other SONE.

The foregoing, of course, is all conjectural at this point, and would call for some dramatic changes to currently accepted ecology theory. Let us proceed with our conjecture, however, in hopes that a vision of value may emerge. Such a raising of status of organisms may result in the reduction and perhaps eventual cessation of culling practices that are currently justified largely out of concern for the well-being of populations. But how might such a marked change in ethical standing occur? Although many factors would undoubtedly come into play, public opinion, attending legislation, and court decisions would likely be necessary components. If such a change occurred, culling might no longer be deemed either ethically or legally defensible. From a positive standpoint, skepticism about SONEs could result in increased conservation emphasis on actions that directly promote the welfare of individual animals, while not detracting from, but rather augmenting, larger-scale projects.

At this point, it seems appropriate to ask: What is a proper ethical basis for grounding such an attempt at reconciling wildlife welfare and ecology?

On what moral ground should such an effort stand? There are several possibilities, each of which I will briefly visit and comment on. These include, in descending order of generality, being, nature, life, sentience, painience, and rationality.

Start off by imagining a situation in which it is possible to conserve or preserve absolutely everything that exists. In a certain sense, this is in fact done automatically through conservation of matter and energy throughout the known universe. To say that everything is conserved, however, is misleading since the structure and function of many things is not conserved automatically because of the entropic forces of chaos and decay. What is more accurate is to say that traces of all physical entities are always in existence. This fact may provide some comfort; however, it does not begin to address the nature of being that such matter and energy participate in. The "furniture" of the world is in constant flux, and we as humans have little power to control this on a grand scale. Moreover, even if possible, it is unclear to what extent such control would be morally or ethically warranted. While every ethic is an ideal, and may be difficult to attain, it should not seem impossible for those who will practice it in the near future. So, active conservation of everything in its present form, even if confined to our planet, would seem a futile task.

Nature, more specifically, is a category in which things (or, from a more process-oriented perspective, systems) either unaltered or little-altered by humans are often placed. Although there are many readings of nature, let us treat the popular view of a domain not significantly, visibly anthropogenically altered. Nature is a favored focus for many environmental ethicists and provides a high ethical bar indeed. However, there are problems with both its meaning and the feasibility of its conservation. While progress is currently being made in rewilding and ecological restoration efforts to remake nature-like states, the question of whether the outcomes of such attempts constitute real or fake nature looms large. Moreover, it remains unclear whether the motivation behind such conservation would be primarily ethical or merely aesthetic.

If moral concern is further restricted to those beings thought alive (Gaia notwithstanding), excluding such natural entities as rocks, soil, water, etc., the set is narrowed considerably. Does this constraint help with feasibility? It seems to, given that humans can survive, and even thrive, on nonliving—indeed, never-living[1]—entities. Fruitarians, for example,

[1] Taking the view that fruits are inessential parts of living entities.

subsist solely on non-vegetative plant matter. However, if attacked by parasites or pathogens, few humans would resist treatment through killing if deemed appropriate. But are they morally permitted to? What should be done when life is pitted against life and no harmless relief can be attained?

Both living itself and the will to live are important and morally relevant traits. Indeed, morality without life as a basis is likely to encounter justificatory difficulty, as in attempts at arguing for the intrinsic value of rivers and rock formations. But the conativism of Schweitzer's ethic of reverence for life (or, more generically, "lifeism") seems problematic as the sole criterion for morality, if for the very fact that the will to live is often decreased because of suffering. What happens when the will to live is challenged by acute pain or disability, when life comes into conflict with will? Does life thereby lose its moral relevance, or become in some way trumped?

Will can be understood in at least two senses: As the basic will that automatically promotes the internal functioning and homeostasis of all organisms, and as the more complex will of "advanced" organisms that is largely under conscious control. It is the latter will to which I am referring in this argument. I do not have a definitive answer to the largely bioethical question of when, if ever, the will to live is justifiably overtaken by sensation of pain or suffering. However, I refuse to discount the negativity of the impact severe pain and suffering have on the lives of individual animals. A term often used in ethics literature to attempt to capture the import of sense experience is sentience—the related term "sentientism" having been defined by Andrew Linzey. Psychologist Richard Ryder eschewed the term in favor of "painism" for three main reasons: "(1) That 'sentientism' might be deemed to refer to any sort of feeling or sensation; (2) that 'sentientism' and 'sentient' were words not popularly understood, whereas 'painism' and 'painient' could be easily grasped and would thus be of greater use politically; and (3) that these words usefully fill some significant gaps in the English language" (Bekoff and Meaney 1998: 270). I agree with Ryder that painism is an improvement over sentientism, since it picks out what sense experience is most morally relevant; however, I hold that preserving the state of living trumps any negative affectation deemed painful, except in the case of the "maximum sufferer" (Ryder 1991). This is because of my belief that, except for in extreme cases, with modern medical technology the hope for a speedy recovery and not only continuation of life, but also eventual attainment of flourishing, seems warranted.

A teleological, Aristotelian approach would comport with this goal of flourishing. Flourishing, however, means something different for different kinds of living things. A utilitarian view on an appropriate goal for life would center on happiness as a quality, in some ways akin to flourishing, that can and should be maximized and summed. However, such a position seems to render the lives of plants and "lower" animals intrinsically meaningless, since they do not appear able to be happy. In addition, the purported summative nature of happiness in utilitarianism allows for suffering to be treated as a state that can be ignored and traded off for a greater amount of good. These are two reasons for why I do not explicitly align my stance with utilitarianism. Another reason is that the term "utilitarian" has an unavoidable connotation of use-value—especially in environmental disciplines.

Lastly, I come to the view that rationality is an appropriate ground for moral standing. This has long been a popular position in Western ethics, with Kant being a major contributor. Most reason-centric positions hold to some version of the claim that nonrational entities have moral value only if they contribute to or detract from the ability of humans to attain their potential as rational beings. However, this brings up the question of the extent to which experience matters morally, since many relatively uneducated and irrational humans do not perceive their cognitive abilities, or lack thereof, as adverse. Ironically, rationality is often thought good mainly because it has utility, whereas pleasure and pain are good and bad intrinsically, in and of themselves. I agree with Bentham that the issue of primary moral significance is whether something can suffer, not whether something can reason.

In response to the formerly posed questions concerning conflicts between life and will, I propose that special attention must be paid to mitigation of those negative feelings that cause the will to live to diminish in living entities who are experiencers of lives, namely, pain and suffering. My position can be categorized as "lifeist," in that it holds to the doctrine of life as the ultimate basis for ethics. I agree with Schweitzer that a truly ethical person "tears no leaf from a tree, plucks no flower, and takes care to crush no insect" (Barsam 2008: 27). However, it should be stated that a life is not necessarily a life—or rather, the kinds of experiences living affords are not evenly distributed among all organisms. The evidence for the possibility of strongly negative experiences seen in all vertebrate animals must be stressed over the lack of evidence for such strongly negative experiences in most other life forms.[2] This admission is part and parcel of

[2] With some notable exceptions, such as octopi.

an active ethic of life-affirmation. Whereas the self-restraint that Schweitzer refers to in relation to not harming plants and insects is an important part of behavior that avoids life-negation, it is unclear whether active help can be given to such entities. Both negative avoidance of causing harm and positive engagement in helping and caring activities, however, are crucial components of an ethic that acknowledges the special relevance of the potential maximum sufferer. So, life is the primary basis for ethics, and pain and suffering must be treated as both bad intrinsically and as hindrances to life. I agree with Linzey, then, "that it is inconsistent to value suffering without also valuing life itself" (2009: 166).

My understanding of the ethical conservation perspective being outlined is that it does not map perfectly onto either current version of individual-based animal ethics (e.g., Regan 1983 or Singer 1975), or a more holistic environmental ethic (e.g., Leopold 1949). It does, however, bear relation to each, and can be seen as a more moderate option. This view is unlike most individualist theories in that it is specifically oriented toward free-living animals. It is like such theories, however, in that it is supportive of actions, policies, and laws that take seriously the welfare of individuals, and unsupportive of actions, policies, and laws that discount the value of individual welfare. Regarding holistic perspectives, there is notable difference in both scalar focus and methods. This, I hold, is because of the already discussed inflation of scale and perception of power common to holistic approaches, as well as the lack of an ethical component within the science of ecology.

Despite attempts to argue for evolutionary and ecological ethics, proof of their appropriateness has heretofore been lacking, except in the rather trivial sense that practicers of ethics and other natural entities are themselves the result of evolutionary and ecological processes. Insofar as an ethic is a subjectively conceived ideal, it must be given to ecology. However, the relevant distinction is not only at the level of science and ethics, fact and value; it is also at the level of entity that is being considered. While a conventional conservation approach emphasizes the integrity and persistence of species and their subunits, populations, animal ethicists tend to focus on organisms, and proponents of environmental ethics look to the large and/or complete systems represented by communities, landscapes, ecosystems, and biomes. Allen and Hoekstra (1992) nicely capture the range of criteria and objects of concern taken by these positions (if not the ethical stances themselves), along with the pluralistic view that they can all exist within a "layer cake" of nature. Such a view, however, does not say much about how different types of actions taken by humans affect the different parts of this layer cake, nor of the appropriateness or desirability of such actions.

The layer cake metaphor suggests that the foci of all the different ecological criteria are composed of the same type of basic material, and that their identity is essentially based on which part of the "cake" one chooses to focus on. The populations of the conservationist, the organisms of the animal ethicist, and the ecosystems of the environmental ethicist are all parallel manifestations of the same fabric of nature, according to such a view. The authors claim that no matter which of the ecological criteria one prefers to use as a researcher, and regardless of how tangible or intangible it seems, one's perspective is characterized by ambiguity. Accordingly, any action taken can be evaluated with equal validity due to the expected effect it would have on any of the criteria one chooses. If this is indeed the case, and conservation biology is properly conceived as applied ecology, then why do conservation biologists seem so concerned about populations and species in particular?

One possible answer to this question is that the unified approach to ecology noted above is a drastic departure from the ecological norm, the truth being that ecologists do take stances on the appropriate scales or objects of ecological concern, and that this is attested to by academic specializations in molecular ecology, landscape ecology, community ecology, etc. If this is the case, then it does hint at a value factor in ecology, in the sense that ecologists are choosing what to study based on some criteria deemed relevant. This choice, however, could be practically, ethically, or aesthetically driven. Although the perspectives taken by many of the defining figures of conservation biology (e.g., Soule or Diamond) are not coextensive with the type of ethical views adhered to by, for example, Leopold, Rolston, and Callicott, the former appear to have a closer affinity to the eco-holistic views than the individualistic views of animal rightists and welfarists. Perhaps the most obvious reason for this is that both conservationists and eco-holists hold that SONEs are the proper objects or levels of conservation concern. But why is this the case? Why do SONEs receive higher regard than, say, organisms, cells, molecules, and atoms? This is not an easy question to answer, perhaps precisely because the answer to it has largely been taken as apparent.

A speculative answer I can offer is the following: SONEs have traditionally been taken as the proper objects of ecological concern because they are, or should be, feasible to preserve and maintain, whereas fates of natural objects of progressively smaller scales become increasingly difficult to control in their vast numbers. Many species, for example, can have large "chunks" taken out of them while apparently keeping their integrity and viability. Scientists, like other professionals, want to have success in their pursuits, and it is more likely for one to have success in conserving a single species than each of 5000 organisms (as the current survival rate of roughly 50% of individuals treated at wildlife rehab centers attests to). The rationale for such a view is largely one of goal-oriented pragmatism. Since each species is valued for a variety of reasons, and because saving a species from extinction should, in principle, be possible, many conservationists assign themselves as guardians of single species or populations. To say that reaching such a goal as the preservation of viable populations is generally possible is not to say that it is easy, or can be done by just anyone. It does, however, raise an interesting, and increasingly relevant, question: As conservationists, how small should we set our sights?

If we back up a bit and take as our point of departure Leopold's land ethic, tracing the development of Western conservation ideals from this highly influential base, some informative trends can be discerned. Although often touted as the founder of wildlife management, Leopold's later work attends more to broad ecological processes than to wildlife per se. Perhaps the most often quoted passage in Leopold's writing, sometimes referred to as the first principle of the land ethic, attests to this big picture view: "A thing is right when it tends to preserve the integrity, stability, and beauty of the biotic community. It is wrong when it tends otherwise" (1949: 224).

At this point, I would like to draw attention to some important questions raised by the position captured in the above statement. What, then, is the sort of thing that can be right or wrong in this context? Does its identity remain obscure because it can be anything? And why is its moral worth defined in terms of its relationship with the biotic community? If the identity of the thing Leopold had in mind was a population or some similar group, it appears to have been posited in terms of the land it occupies. A population, after all, must have populated an area, and this framing of a population as "that which has populated" is often ignored. So, has land- (and therefore landscape-) centrism, in terms of which what populates terrestrially has been framed, created an implicit bias in modern conservation toward the occupied rather than the occupant?

Community is a term seen very often in Leopold's work and appears to go hand-in-glove with his frequent mention of the land, insofar as the land is that which ultimately facilitates community function through time. This is not to say that Leopold was unconcerned with, for example, water or air quality, but his special focus on the land community shows a great amount of emphasis put on the significance of terrestrial existence. We see in the land ethic a placement of value on large-scale ecological processes and functions, often with a relative disvalue placed on individual lives.

Many contemporary environmental ethicists have basically picked up where Leopold left off—Callicott, for example, with the unambiguously titled *In Defense of the Land Ethic* (1989). Such a holistic stance taken to its logical conclusion can be observed in Naess' deep ecology movement (1973), as well as in Lovelock's Gaia hypothesis (2000). In addition, many feminist environmental philosophers agree with certain elements of the land ethic. This includes, but is not limited to, a "notion of the self as a relational, ecological being who is a member of the larger biotic (living, organic, ecological) community" (Warren 2015).

Conservation biology, while interdisciplinary in scope and ideally inclusive of many different perspectives and ecological criteria, has remained focused on the next rung down on the ecological ladder, so to speak, from the biotic community. The stepping down to the population level that has reputedly been accomplished could perhaps be more accurately portrayed as a leap of faith, since the proof of this "rung's" existence, as we have discussed, has been shown to be far from conclusive. While the community concept includes multiple species, and even abiotic entities, conservation efforts of the recent past have largely been fixed on what needs to be done to keep populations (and, ultimately, species) from either going extinct or getting too small. A hard-and-fast distinction between focusing on communities and populations, or other comparison of ecological levels, cannot always be made, but downscaling of efforts is often a reactive need when dealing with species in peril. Although implementing global greenhouse gas emission reduction and protecting Gila trout (*Oncorhynchus gilae*) from a proposed development, for example, may be complementary activities, the latter may have to be enacted in a timelier fashion to achieve the desired outcome.

While downscaling and localizing compared to more holistic initiatives may be seen as an improvement by some for purposes of saving individual species, Leopold's framing of conservation biology's midlevel objects of concern as those entities which rightfully or wrongfully exist on a particular landscape continues to come through. Aside from the previously discussed metaphysical issues about what a population is composed of other than organisms, a potential logical fallacy (more specifically, an ecological fallacy) may obtain with ethical conclusions drawn about where a population happens to be located at a particular point in time. The common form of this fallacy, committed by trying to derive an "ought" from an "is," a prescription from a description, without other justification, is often referred to as the naturalistic fallacy.[3] I would like to suggest, along with Lockwood and Latchininsky (2008), a special case of this kind of ethical jump that more accurately pertains to attempts at deriving an ought from a "was." This faulty line of reasoning has been referred to as the conservationist fallacy. It is seen as particular to conservation because conservation is marked by its consistent focus on returning to a past situation. Hopefully, such a fallacy can be purged from association with conservation before it becomes inextricably linked to the discipline. To gain a clearer view of the problems associated with trying to derive an ought from a was, let us take a closer look at what such a jump might entail. First, however, we must try to gain a better sense of what nature can mean for us.

There are many versions of or takes on the more well-known naturalistic fallacy, the most notable stemming from the work of David Hume (1739) and G. E. Moore (1993). These interpretations vary greatly according to one's understanding of the meaning of naturalistic and, indeed, nature. Although the nature cited in the naturalistic fallacy is often understood as referring generally to "the way things are," for our context it is taken to mean something like the more common understanding of nature as a domain unaltered by human influence. As has been touched on previously and will be developed further, both the extent to which nature has been altered and how this alteration affects conservation practice develop into important questions.

In like manner, the form the conservationist fallacy takes will vary by the interpretation of conservation taken. Although a precise formulation of the conservationist fallacy has, perhaps thankfully for conservationists, yet to be given, the problem is apparently with how we arrive at conclusions about how nature should be acted in and upon. It is, then, not only about a desirable state of nature but also about how a state that existed in the past, insofar as it can be brought about again, should be brought about again. A central issue with such an approach is the potential for essentially arbitrary aesthetic values derived from a desired past time to be passed off as ethical imperatives.

[3] A distinction is sometimes made between the naturalistic fallacy and the "is-ought problem," but focusing on this distinction is not important for the present argument.

We can attempt to gain enhanced knowledge of nature by examining its placement at either of two poles. At one extreme, there is the belief that all is natural, or at least partakes to an extent in a past naturalness, since every piece of reality can in some way be derived from or traced back to a prior state unaltered by humanity. Central to such a view of nature is the idea of humanity as inherently natural all the way down, such that whatever humans affect, either directly or indirectly, remains in some sense natural (Callicott 1996). Humans, after all, are evolved organisms, and differ from the rest of nature, if at all, by way of degree and not kind. Commitment of the naturalistic fallacy, on this account, would seem to be difficult to avoid, given the putative naturalness of everything and, therefore, assured justificatory reference to "the way things are."

At the other extreme is the statement that true nature no longer exists (McKibben 1989). It existed at one time, but given the ubiquitous effect of humans on the environment, nature has been eternally purged from the world's furnishings. A foundational assumption of this view is that humans are different in kind from other life forms and can therefore cause different kinds of outcomes to occur. Moreover, this human influence is not only direct and obvious, but also indirect, hidden, and mediated through technology. When we use technology to create artifacts, we are doing something new under the sun. According to this view, an object or state that is somehow altered either directly by humans or indirectly by way of technological objects thereby loses its quality of naturalness.

Such a perspective suggests a very severe contamination of the world by humans. Human influence is thought to have affected the entire biosphere, such that, if every object itself is not completely unnatural, the climate surrounding every object on earth is. It is unclear just how, in this world, a naturalistic fallacy in our sense could even occur. Since we cannot knowingly refer to a nature that presently exists, we cannot derive an ought from a natural is. The best we can do if we want something natural, on such a view, is attempt a remake modeled after our most complete knowledge of a prior state.

A more moderate, perhaps common-sense alternative view states that, while the impact of humans on the earth is widespread and pervasive, it is incomplete. Those entities that are clearly modified by humans are considered technologies or technological artifacts, while those that do not owe their existence, form, or function primarily to human intent or design are considered natural objects. Some examples in the former category are computers, buildings, and automobiles, while examples in the latter category are wild animals, plants, and rocks. The moderate view of nature takes seriously the role of human intent in determining what is nature and what is not.

According to this interpretation of nature and the naturalistic fallacy, it would be wrong to, for example, conclude that predation is good and something that ought to be engaged in because it is the case that wild carnivores hunt and kill prey. Another instance of such an error might be the attribution of goodness to a specific river simply because it has not been significantly altered by development. This conception of natural goodness seems closely linked to the troubled notion of intrinsic value, which has been used by many environmental ethicists to attempt to ground and defend the preservation of nature.

Sandler makes the distinction between subjective and objective intrinsic value. Subjective intrinsic value is conferred by humans on natural objects for what they are rather than what they can bring about, while objective intrinsic value is thought to inhere in natural objects independently of anyone's opinion or judgment (2012). While this is a welcome distinction at one level, at another level it is not particularly helpful, since both positions reduce to beliefs about what ought to be the case that are ultimately derived from beliefs about what is the case. As a foundation for an ethical theory, intrinsic value is found wanting.

Now that we have considered some of the ways in which the nature of the naturalistic fallacy can be conceived, we can begin to grasp the so-called conservationist fallacy a bit better. Notice that, properly speaking, the is-ought problem can be understood in nearly every case as a was-ought problem, since the present is continually receding into the past. It is surprising that this point has hardly been discussed. In any situation where a jump is occurring from the way things are to the way things should be, if sliced down to the smallest time step, an assumption is in fact being made about the way things should be from the way things were. The characterization of this type of logical error as pertaining solely to conservation, or of being the main problem of conservation, may be far too narrow. It at least seems likely, however, that this fallacy has the potential to negatively affect environmental disciplines if sufficient other support is not provided. I will attempt to provide insight on this and other trying aspects of conservation biology by briefly looking at three major issues commonly addressed in conservation theory and practice. In addition to outlining the conventional conservation perspective on each issue, for purposes of contrast and clarification I will give examples of the more organism-centric view held in wildlife rehab.

First is the issue of invasiveness. Conservation biologists are nearly universally concerned with protection and preservation of species considered native to a certain region. The argument often goes like this in North America: If a species arrived pre-European arrival, it is native. If a species arrived post-European arrival, it is nonnative and likely invasive, should not be protected or aided, and should probably be exterminated—unless, like the ring-necked pheasant (*Phasianus colchicus*), for example, it is deemed valuable for other reasons such as hunting. However, many (but certainly not all) wildlife rehabbers make no discrimination according to native or invasive status, under the conviction that these categories are basically arbitrary, and with the belief that, regardless of when or how it arrived where it is, an animal that is ill, injured, or orphaned deserves proper treatment. After all, humans probably provided the opportunity for it to get here, and we have likely, directly or indirectly, caused the problems it is now having.

This is an area of current conflict between the disciplines and is likely to prove difficult to resolve. It also has the potential to open paths to commitment of the conservationist fallacy, because of the often-unjustified favor shown to a particular evolutionary time slice. To be fair, the danger does exist that a nonnative species will outcompete or prey heavily upon a species that has been present for a longer amount of time in a particular area. One of the main examples of this is the Nile perch (*Lates niloticus*) decimating cichlids in Africa. However, there is also the possibility that competition, if it occurs, will lead to enhanced selection for fitness in the long run, thereby aiding both species. Moreover, use of contraceptive methods is on the rise, potentially helping to keep nonnative species humanely in check.

Moral consideration of nativity and invasiveness brings up the difficult question of fault and duty. Because some species have arrived at their current location due to human actions, do we now have the perhaps unwanted duty to treat them as we would so-called native species? Should ethologist Marc Bekoff's rule for our interactions with animals, "treat them better or leave them alone" (2010: 1), be extended to nonnative species? Notice that this treatment would not lessen responsibility for prevention of such movements occurring in the first place in the future.

The issue of native versus nonnative status is among the most hotly contested in the conservation/rehab debate and has led to conflict between conservationists and rehabbers (Phil Jenni, pers. comm.). It is likely more intractable than the previously discussed issue of rarity, since it is associated with a higher degree of certainty once an agreement as to a definition is made. It may be known with a higher degree of certainty whether members of a species exist in a given region within an agreed upon broad time frame than the precise level of endangerment within a much narrower time frame may be known. If precolonial presence is agreed upon as a sufficient condition for a species to be native, and this condition is observed, then the pressure for conventional conservation remains strong among those who see these criteria as relevant. The case, however, is far from airtight, since agreement that native means "became established pre-colonially" does not render the concepts of native or invasive non-arbitrary. Many seem to want "invasive" to be synonymous with "ecosystem-destructive," and in some cases, such as in aquatic ecosystems, this equivalence seems valid. However, Schlaepfer et al. outline the potential conservation value of nonnative species (2011). And even if some invasives are destructive, is this sort of destruction akin to the destruction humans have wreaked and continue to wreak on ecosystems, of their own free will, through various technological means?

Conservationists, then, should discern and acknowledge the potential for short-sightedness and arbitrariness in native designations. This might lead to a more serious examination of the possibility of nonnative species offering good selective pressure and other benefits in some cases, and perhaps even the adoption of a fairer "innocent until proven guilty" approach. At the least, attempts should be made to place more emphasis on investigation of the feasibility of humane, non-lethal methods. Compassionate conservation, for example, "means consideration for the interests of individuals from introduced species, often labeled as invasive by conservation scientists who disregard the pain and suffering associated with relocation or eradication of those individuals in order to preserve or restore biodiversity and native ecosystems" (Lynn et al. 2025: 5). Although economic and technological constraints may seem to hinder immediate adoption of nonlethal approaches, in many cases voluntary, low-tech strategies are sufficient to significantly reduce numbers. An example of this is removal of feral pigeon (*Columba livia*) eggs from nests by volunteers in Bern, Switzerland (Peggy Naylor, pers. comm.).

Second is the issue of restoration. Among conservationists, two general schools of thought have emerged: Preservationist, where large tracts of habitat, entire populations, etc. are protected and human intervention is kept to a minimum; and restorationist, where attempts are made to bring degraded and otherwise suboptimal environmental states closer to a desired functional level through intensive human intervention. Central to the preservationist approach is the intuitive judgment that it is better to preserve large entities intact (landscapes, forests, biomes) than to restore them. Preservation has historically been quite successful in the West; however, restoration of habitat and plant assemblages is increasingly called for.

Such restorative efforts are increasingly being referred to as "rewilding." Although the term has its origins in North American conservation, it is increasingly being employed to denote intervention efforts in areas (such as Europe) with high human population densities. Rewilding is a type of restoration practice that explicitly acknowledges the corruption of nature by humans and technology and seeks to remedy the resulting imbalance. This is done through practices such as "restoration of natural ecosystem processes through improved connectivity, species restoration and/or creation of core wilderness areas while minimizing human management to create self-regulating and self-sustaining ecosystems" (Massenberg et al. 2023: 47).

Since much of rewilding is undertaken at the ecosystem level, it is generally conceived of at a large spatial scale. Wildlife rehab aligns with the overall restorationist program but is much smaller scale in its focus on individual organisms. Rehab is also preservationist, however, since it aims to preserve lives. Wildlife rehab is essentially individual animal restoration, but it differs from restoration of plant assemblages in that plant restoration often focuses on replacement rather than repair of biotic entities.

As with many of the preceding apparent oppositions, this dichotomy is largely false. It seems that preservation and restoration can both be practiced, and often simultaneously, as in rehab cases where lives can be preserved and abilities restored. Since we have arguably been historically more successful with preserving large tracts of lands than in restoring impaired ones, and because restoration and rehab typically happen on smaller scales than ecological preservation, there are many more opportunities for them. Moreover, this is increasingly becoming the case as preservation efforts are being undone by diffuse and largely unpredictable means—most notably, climate change (Schlaepfer and Lawler 2023).

Restoration and rehab will likely become increasingly recognized as practices of import for places, like most places, where preservation has been incomplete or reversed by factors such as climate change. This is not to deny the adaptability and resiliency of many organisms in the face of changes to their habitat, but to acknowledge the likelihood that even with aggressive mitigation measures in place, many places on earth will simply never be the same again after our interference. Notice that we have not yet been able to rid ourselves of the danger of committing the conservationist fallacy, however, since restoration efforts are often justified by perceptions of invasiveness and desires for a return to the past.

To what extent are the preservation and restoration "schools" in conflict? Although the answer is not entirely clear, the truth is that the practice of restoration has been fraught with conflict. For example, as Gobster, Hull, and others have discussed, in areas where restoration activities have harmed animals and removed trees in favor of grasslands, conflict has at times been fierce (Gobster and Hull 2000). Central to this form of restoration is the concept and practice of replacement—the notion that parts of an ecosystem, whether plant or animal, are merely good or bad pieces of a larger whole, and if we take them out better ones will either come in or be put in. The sense in which wildlife rehab restores, however, is different in that it is in a sense a truer form of restoration, one without replacement of organisms. Different versions of restoration aside, it should be acknowledged that rehab is in fact a form of both small-scale preservation and restoration. It is, in essence, preservation and restoration of individual lives. Let us then consider acknowledging the identity of restoration and rehab as the other side of the conservation coin—what is needed when preservation is incomplete.

Third is the issue of rarity. Conservation biology has been focused on preventing extinctions of populations and species; therefore, its first concern has been members of endangered, scarce, or rare groups, or those kinds thought to be at risk for extinction. Practitioners of wildlife rehab, however, less often attempt to make explicit judgments according to perceived abundance or commonness, and more frequently treat organisms of common species. Exceptions to this exist, however, in the form of specialists such as endangered species veterinarians.

In addition to the welfare benefits that accrue to treated animals, there is the potential for using such commonly treated animals as eastern gray squirrels (*Sciurus carolinensis*) and eastern cottontails (*Sylvilagus floridanus*) as models for similar endangered or data deficient species, like, for example, the Arizona gray squirrel (*Sciurus arizonensis*) or Manzano Mountain cottontail (*Sylvilagus cognatus*). Many people value the members of common and seemingly mundane species more than those less common for the simple reason that they interact with them more often, and because of this know and can relate to them better. Because of this knowledge, many people are willing to donate to causes such as wildlife rehab. So, intrinsic worth aside, the cultural value of common species may trump that of rare species.

There are some promising approaches to dissolving the treatment gap between the rare and the common without ignoring what science suggests. However, the concept of viable population size is essentially a guideline, and while it may tell us something of the effect of maximizing genetic diversity to, for example, avoid inbreeding depression, it is likely the case that it is natural for different kinds and sizes of animals to have different optimal population sizes, if such an optimality in fact exists. Following from this uncertainty, one approach is to adopt ethical principles that are less focused on group membership, and more responsive to the concreteness of individual need.

Theory begins to look quite different when one raises the status of ethics in relation to science, however. We will move now into a discussion of the particularly ethical characteristics of conservation and rehab, in the hopes that the possibility of alternative, compatible readings of each of the issues discussed has been seen. Environmental ethicist J. Baird Callicott, in an early attempt to show the incompatibility of ecological holism and individual animal ethics, has posed a problem in the form of a triangle (1980; see Fig. 7.1).

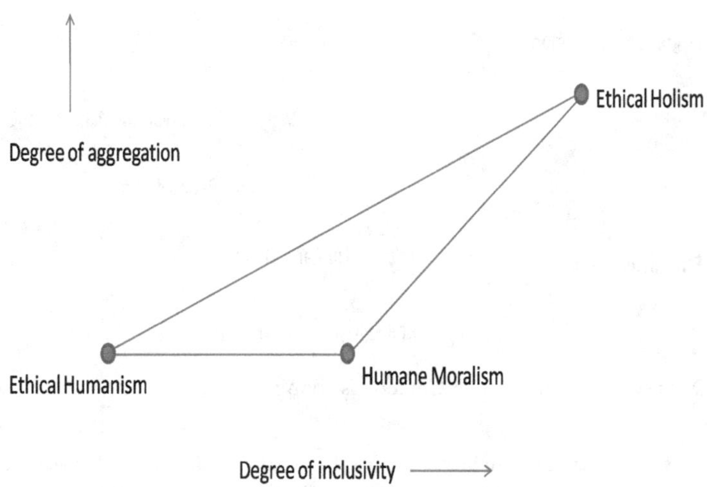

Fig. 7.1 Callicott's triangle

The x axis of the triangle progresses from left to right in degree of inclusivity of different kinds of entities, and the y axis progresses from bottom to top in degree of aggregation. He treats the three broad ethical stances of humanism, or a human-centered ethic; holism, or an all-inclusive, large-scale ecological ethic; and humane moralism, or an ethic centered on the humane treatment of animals. He contends, in essence, that holism and humane moralism are in conflict, since they are on distinct planes and seem to advocate for treatment of different entities. Although Callicott goes on to soften his initial position (1988, 1998), it remains influential and typifies a formal version of a common ethical split between conservationists and rehabbers. The differing theoretical foundations of science and ethics are part of the reason, I think, for Callicott's belief that the relationship between ethical humanism, humane moralism, and environmental ethics is not straightforwardly linear.

In attempting a solution to this problem, I found that, if mapped in terms of inclusivity and aggregation, conservation and traditional wildlife management,[4] on the one hand, and wildlife rehab, on the other, are not so dissimilar as I had believed (see Fig. 7.2). If one shifts down in scale from the extreme holistic focus on the entire earth to a focus on species or

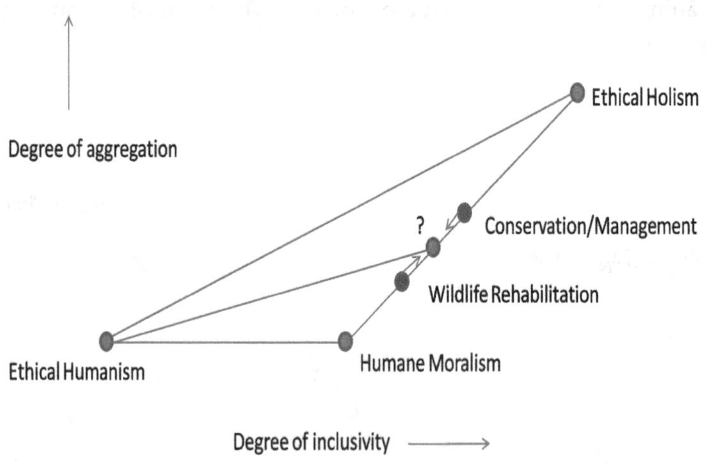

Fig. 7.2 Potential solution to Callicott's triangle

[4] I acknowledge that there are both theoretical and practical differences between conservation biology and wildlife management. However, their concurrent focus on the importance of SONEs warrants similar treatment for the purpose of this argument.

populations, one arrives at a point near the center of the line. In like manner, if one moves from domestic animal care to wildlife rehab, one nears the center of the line. What is left is a relatively small distance between conservation and rehab, within which efforts can be made to increase linearity, showing agreement and compatibility. As the disciplines move closer together, a robust and more unified challenge to the dated ethical position of humanism may yet emerge. The question arises: Why not simply stay at the higher right side of the triangle and insist on the total preservation of everything natural on earth? The answer is that it is impractical—in fact, impossible—to preserve all of nature in a vacuum, and that this impossibility is not lost on those who would remain staunchly anthropocentric. It is also the case, however, that we can do better as conservationists by setting the bar higher in terms of those entities that we deem worthy of protection and care.

Most Western ethical theories have started with the individual. This is because the individual is the locus of reason, awareness, and sentience. These attributes, broadly, are the qualities that have been most often understood as morally relevant by modern and contemporary Western ethicists. The two main broad schools of modern ethics are deontology (e.g.. Kantianism [Kant 1993]) and consequentialism (e.g., utilitarianism [Mill 1864]), the former focusing on duty, the latter on consequences. Historically, deontologists have only recognized duty to rational beings, and it has been thought that humans are the only species with reason. We have seen this assumption being challenged, however, by studies (e.g., Benz-Schwarzburg and Knight 2011) indicating that many species seem to possess language, theory of mind, culture, and other attributes only thought possible for humans. Regan's influential position on animal rights stems from an argument for the recognition of duties owed to rights holders, who have these rights in virtue of expressing certain "higher" qualities (1983). Singer's animal liberation ethic also begins with the individual, but "lowers the bar" to those species that merely express signs of experiencing pain or suffering, believing this capacity to be the baseline for having moral consideration (1975).

The positions espoused by Regan and Singer have grown up alongside conservation biology, and both have developed during roughly the same time with their paths hardly ever meeting. As Callicott points out, individualistic viewpoints have traditionally dealt with issues of domesticity, whereas ecological (and conservation) traditions have taken as their major subject the wild and natural. One of the reasons for this is the received Christian view of the separation of humanity from the rest of nature that many ethical theories stem from. Unless natural entities have been invited into human society, as in Midgley's notion of the "mixed community" (1983), they have historically been considered as wholly other. On the relatively rare occasion that animal welfarists and rightists have discussed "wild" animals, it has usually been in the context of those that have been forced into the human community as zoo specimens, or the prohibition of actions such as hunting and trapping that are seen as negative. But what, if any, positive duties do we have toward animals who are not explicitly recognized as members of this mixed community (Donaldson and Kymlicka 2011: 49)? Wildlife rehab suggests a new, positive approach that has the power to fundamentally transform how conservation is conceived of and applied.

The growing compassionate conservation movement acknowledges both the tension between individual animal welfare and large-scale conservation, and the possibility of reconciliation. While an "overemphasis on collectives" is acknowledged in conventional conservation practice, "care for individual animals and plants can co-exist with care for ecosystems" (Lynn et al. 2025: 4). Even within a more conventional conservation framework, the value of wildlife rehab is beginning to be noticed. A 2024 study, for example, attempted to identify both direct and indirect benefits of wildlife rehab for conservation by way of a literature review. Although the review itself was incomplete, enough data were collected to support a call for enhanced collaboration and an increase in research (Mladonicky et al. 2024). Most importantly, this study attests to the fact that a professional bridge is being built between two formerly conflicting but potentially complementary disciplines.

A major issue that comes up with rehabbing is the question of the right (and, by extension, potential duty) to intervene in wildlife affairs. We may have a "negative" duty to do no harm, but this does not necessarily imply that we have a matching "positive" duty to help. This can be framed as a potential conflict over what takes precedence—preservation of the "wildness" of a wild animal or treatment of pain, suffering, or other problems it encounters as an organism. Moreover, if humans are deemed to have duties toward individual wild animals, the question then becomes: Who has these duties? Individual rehabbers and/or rehab centers, governmental agencies, or both? The former usually rely on volunteers, and currently do the lion's share of rehab work, whereas the latter have paid employees and often financially gain from wildlife killing programs. Why should those who are rewarded the least (and often have a lesser amount of scientific training) do the most "on the ground" work?

The earlier analysis of wolves and killing centered on a suite of well-known, straightforwardly life-taking practices. A subtler, indirect form of harm is often caused when the good of some organisms is prioritized over that of others because of their perceived group representation. Such a view has been codified in such legislation as the ESA, where members of species deemed in danger of extinction enjoy special provisions not extended to species lacking adequate proof of low numbers. From a conventional conservation viewpoint, the justification for such preferential treatment appears immediately valid. If, for example, the Tellico Dam is built, number and range estimates indicate that the kind known as the snail darter (*Percina tanasi*) may significantly decline or cease to exist, whereas, even if thousands of white suckers (*Catostomus commersonii*) are killed by a similar construction project on a different river with no snail darters or other species defined as endangered, this will not harm the white sucker species, since the local population has been deemed abundant or over-abundant. So, the latter project goes forward easily, while the former is stalled. After all, although numerous deaths are a predictable effect of such a project, no fish is singled out and intentionally killed; the deaths are an "accidental" consequence of a development initiative.

While wildlife rehab undoubtedly makes a difference for those individual animals who are treated by its practitioners and methods, and, unlike conventional conservation biology, is often unbiased in relation to background conditions of species like rarity, nativity, and assumed functional role, it might also be able to meet conventional conservation on common ground. This is the "ground" of great uncertainty, where sufficient data have not been collected to (yet) deem a species worthy of protected status, but where such data could likely be found in the future. Ironically, such lack of data itself is currently a hindrance to listing and protection, and in this way a lack of evidence can count as evidence against protection for the mere fact that scientists have not prioritized studying a particular species. Unfortunately, explicit and implicit pleas for help all too often fall on deaf ears because of irrelevant and often uncontrollable contingencies, such as species, race, sex, or religion, and the prejudice that often attends contact with these categories.

What I am referring to here is the problem of data deficiency, and this problem brings up a host of questions about whether the process of deeming animals worthy of aid is as fair and unbiased as it should be. The ethical bar for conservation needs to be raised, while the evidentiary bar is lowered. This does not mean that biologists should stop counting, drop everything, and focus on rehabbing, or that every animal we observe that we think might need help should be assisted. However, it is in part a critique of the soft view of populations, insofar as lack of evidence regarding numbers is a hindrance to active conservation.

If I am reading the history of conservation correctly, I understand the drive of the movement to be more proactive than reactive. One potential major problem with wildlife rehab is that it tends more toward reactivity than proactivity. An eye toward being proactive would seem to entail accepting the precautionary principle, where uncertainty is acknowledged and caution taken to plan. However, an irony of current conservation practice is that, to get to the planning stage, we are often constrained to wait until sufficient data of a species' endangerment is accumulated. When the population status of a particular type of organism is unclear, should we wait for more data before acting? What if, during this waiting period, an entire population or species goes extinct? This, the problem of data deficiency, is both a major issue and a place where conservation and wildlife rehab can converge in the future. An example perhaps especially germane to rehabbers is the case of the Arizona gray squirrel, a tree squirrel species in the southwestern United States listed by the International Union for Conservation of Nature and Natural Resources (IUCN) as "data deficient" (2025).

An under-recognized yet hopeful aspect of wildlife rehab is the potential for refinement of rehab methods for potentially endangered species by using similar, more common species as models. For example, many rehabbers have large amounts of experience working with other North American tree squirrel species, and this may prove helpful to Arizona gray squirrels. For all we know right now, the life of every individual of data-deficient species may be crucial for the survival of the species. Do we wait for more data to come in, or focus rehab efforts now, based on experience and knowledge gained from working with more common, related species?

References

Allen, T. F. H., and T. W. Hoekstra. 1992. *Toward a Unified Ecology*. New York: Columbia University Press.

Barsam, A. P. 2008. *Reverence for Life: Albert Schweitzer's Great Contribution to Ethical Thought*. Oxford: Oxford University Press.

Bekoff, M. 2010. *The Animal Manifesto: Six Reasons for Expanding our Compassion Footprint*. Novato: New World Library.

Bekoff, M., and C. A. Meaney, eds. 1998. *Encyclopedia of Animal Rights and Animal Welfare*. Westport: Greenwood Press.

Benz-Schwarzburg, J., and A. Knight. 2011. Cognitive Relatives Yet Moral Strangers? *Journal of Animal Ethics* 1 (1): 9–36.

Callicott, J. B. 1980. Animal Liberation: A Triangular Affair. *Environmental Ethics* 2:311–338.

Callicott, J. B. 1988. Animal Liberation and Environmental Ethics: Back Together Again. *Between the Species* 4:163–169.

Callicott, J. B. 1989. *In Defense of the Land Ethic: Essays in Environmental Philosophy*. Albany: State University of New York Press.

Callicott, J. B. 1996. Do Deconstructive Ecology and Sociobiology Undermine Leopold's Land Ethic? *Environmental Ethics* 18 (4): 353–372.

Callicott, J. B. 1998. "Back Together Again" Again. *Environmental Values* 7:461–475.

Donaldson, S., and W. Kymlicka. 2011. *Zoopolis: A Political Theory of Animal Rights*. Spain: Oxford University Press.

Gobster, P. H., and R. B. Hull, eds. 2000. *Restoring Nature: Perspectives from the Social Sciences and Humanities*. Washington, DC: Island Press.

Hume, D. 1739. *A Treatise of Human Nature: Being an Attempt to Introduce the Experimental Method of Reasoning into Moral Subjects*. London: John Noon.

International Union for Conservation of Nature and Natural Resources. 2025. *IUCN Red List of Threatened Species*, Version 2025-1. IUCN Red List of Threatened Species. https://www.iucnredlist.org/en. Accessed 14 June 2025.

Kant, I. 1993. *Grounding for the Metaphysics of Morals: With, On a Supposed Right to Lie Because of Philanthropic Concerns* (3rd ed.). Trans. J. W. Ellington. Indianapolis: Hackett Publishing.

Leopold, A. 1949. *A Sand County Almanac and Sketches Here and There*. New York: Oxford University Press.

Linzey, A. 2009. *Why Animal Suffering Matters: Philosophy, Theology, and Practical Ethics*. Oxford: Oxford University Press.

Lockwood, J. A., and A. V. Latchininsky. 2008. Philosophical Justifications for the Extirpation of Non-indigenous Species: The Case of the Grasshopper *Schistocerca nitens* (Orthoptera) on the Island of Nihoa, Hawaii. *Journal of Insect Conservation* 12 (3–4): 429–445.

Lovelock, J. 2000. *Gaia: A New Look at Life on Earth*. 3rd ed. Oxford: Oxford University Press.

Lynn, W. S., L. Baker, W. T. Borrie, A. P. A. Cardilini, S. M. Alexander, S. Coghlan, P. Cryer, G. T. Bonsen, T. T. Derham, O. Keynan, C. M. Reed, S. Riley, E. A. Ryan, F. J. Santiago-Avila, K. Walker, A. E. Weiss, and N. Xenakis. 2025. Compassionate Conservation in Practice: A Values-Driven, Interdisciplinary, Pluralistic, and Deliberative Community. *Biological Conservation* 303:111002.

Massenberg, J. R., J. Schiller, and C. Schröter-Schlaack. 2023. Towards a Holistic Approach to Rewilding in Cultural Landscapes. *People and Nature* 5 (1): 45–56.

McKibben, B. 1989. *The End of Nature*. New York: Random House.

Midgley, M. 1983. *Animals and Why They Matter*. Athens: University of Georgia Press.

Mill, J. S. 1864. *Utilitarianism: Reprinted from "Fraser's Magazine"*. 2nd ed. London: Longman, Green, Longman, Roberts, and Green.

Mladonicky, J., M. Sirolli, M. Willette, and J. Ponder. 2024. *U.S. Fish and Wildlife Service Migratory Bird Program*. Evaluation of Current Scientific Literature on the Impact of Wildlife Rehabilitation on Conservation, Documentation of Knowledge Gaps and Identification of Future Opportunities. https://www.fws.gov/media/evaluation-current-scientific-literature-impact-wildlife-rehabilitation-conservation. Accessed 1 March 2025.

Moore, G. E. 1993. *Principia Ethica*. 2nd ed. Cambridge: Cambridge University Press.

Naess, A. 1973. The Shallow and the Deep, Long-Range Ecology Movement: A Summary. *Inquiry* 16:95–100.

Regan, T. 1983. *The Case for Animal Rights*. Berkeley: University of California Press.

Ryder, R. D. 1991. *Painism: Ethics, Animal Rights, and Environmentalism*. Cardiff: University of Wales.

Sandler, R. 2012. Intrinsic Value, Ecology, and Conservation. *Nature Education Knowledge* 3 (10): 4.

Schlaepfer, M. A., and J. J. Lawler. 2023. Conserving Biodiversity in the Face of Rapid Climate Change Requires a Shift in Priorities. *Wiley Interdisciplinary Reviews: Climate Change* 14 (1): e798.

Schlaepfer, M. A., D. F. Sax, and J. D. Olden. 2011. The Potential Conservation Value of Non-native Species. *Conservation Biology* 25 (3): 428–437.

Singer, P. 1975. *Animal Liberation*. New York: Avon.

Warren, K. J. 2015. Feminist Environmental Philosophy. In *Stanford Encyclopedia of Philosophy*. https://plato.stanford.edu/entries/feminism-environmental/. Accessed 10 December 2024.

CHAPTER 8

The Alien Encounter Thought Experiment

Abstract While much of the foregoing has been about conceptual issues in conservation and what ought not be done as a conservationist, I would like to continue transitioning into a more positive approach out of the belief that appropriate activity is a critical part of biotic conservation. It is easy to take an ethical approach that dwells on the negatives of past and present abuses, whereby one is effectively frozen. However, if conservation is to become a discipline where service is taken seriously—whereby, as the name suggests, we truly serve together—a move must be made, where possible, from inaction to compassionate action. The following set of hypothetical scenarios are meant to help one mentally simulate variants of a situation in which a crucial decision must be made to act or not act on behalf of a creature. These scenarios, although fictional, are in certain respects not unlike the seemingly random encounters individuals have where they are forced to decide whether to rescue or not rescue an animal apparently in need. Some key aspects of the scenarios, however, mark them as departures from the norm, and help demonstrate the current confusion of motivations and reasons for conservation. I have chosen to construct them as thought experiments because of the privacy that often shrouds people's encounters with wild animals, coupled with an attendant lack of accountability for those animals' vital interests.

L. B. DeVaney, *Animal Conservation Ethics and the Population Problem*, The Palgrave Macmillan Animal Ethics Series,
https://doi.org/10.1007/978-3-031-91891-9_8

Keywords Thought experiment • Scenario • Motivation • Nonnative • Value

While much of the foregoing has been about conceptual issues in conservation and what ought not be done as a conservationist, I would like to continue transitioning into a more positive approach out of the belief that appropriate activity is a critical part of biotic conservation. It is easy to take an ethical approach that dwells on the negatives of past and present abuses, whereby one is effectively frozen. However, if conservation is to become a discipline where service is taken seriously—whereby, as the name suggests, we truly serve together—a move must be made, where possible, from inaction to compassionate action. The following set of hypothetical scenarios are meant to help one mentally simulate variants of a situation in which a crucial decision must be made to act or not act on behalf of a creature. These scenarios, although fictional, are in certain respects not unlike the seemingly random encounters individuals have where they are forced to decide whether to rescue or not rescue an animal apparently in need. Some key aspects of the scenarios, however, mark them as departures from the norm, and help demonstrate the current confusion of motivations and reasons for conservation. I have chosen to construct them as thought experiments because of the privacy that often shrouds people's encounters with wild animals, coupled with an attendant lack of accountability for those animals' vital interests.

Imagine the following scenario. A biologist is in the field, performing research on a remnant parcel of native prairie. It is the end of a hard day's work, and she is tired. The biologist hikes back to the parking lot where her car is located. She is about to reach her car when, at the top of a small rise beyond, she spots something moving. She is curious and decides to get a closer look. As she approaches the unknown being, she realizes that she cannot seem to identify it. The closer she gets, the stranger it seems. It is clearly alive and flapping around and appears injured. When she finally reaches it, she is amazed. It appears small and somewhat birdlike but is covered in scales and oozing a green liquid. Although the biologist is an experienced naturalist, she has never seen or read of such an animal. "This could be an important discovery!" she thinks to herself, and despite her apprehension she picks the creature up, carries it to her car, and brings it back to a colleague's laboratory at a nearby university to be studied.

Now, imagine this similar but quite different scenario. The same biologist is in the field, performing research on a remnant parcel of native prairie. It is the end of a hard day's work, and she is tired. The biologist hikes back to the parking lot where her car is located. She is about to reach her car when, at the top of a small rise beyond, she spots something moving. She is curious and decides to get a closer look. As she approaches the unknown being, she recognizes that it is a bird; but she is unsure as of yet of its species. It is clearly alive and flapping around and appears injured. An experienced naturalist, the biologist wonders what species it is and moves closer still. While yet a short distance away, she identifies it as a common starling (*Sturnus vulgaris*). She understands starlings are a common, non-native species. Her curiosity assuaged, she turns back and heads to her car.

What do we take away from these two hypothetical scenarios? Although they are fictitious, neither is hard to imagine (although the first is implausible). In addition, the way in which the biologist responds in each situation is believable, and in accord with intuition about how one might respond given such an encounter. While the scenarios appear relatively simple and straightforward, they have the potential to help us conceive of likely motivations for actions toward different wild creatures. Perhaps the most noteworthy difference between the two scenarios is that, in the first example, the biologist captures the creature and transports it to the university, whereas in the second example, she leaves the animal where she finds it and does not intervene. Why the marked difference in responses? In each case, the creature appears injured and would likely benefit from treatment or care of some kind. One possibility for the scientist not helping in the second case is that she is afraid of being harmed by the creature in some way. This, however, would seem unlikely, given the familiarity of the scientist with its species, as well as the nonaggressive behavior typically shown by starlings. Moreover, if fear were a hindrance to capturing the creature in either case, one would expect it to be so in the first rather than the second, given that the ecologist does not know what kind of entity it is; but it is not so.

Another possibility is that the ecologist is biased in some way, or prefers certain categories of organisms to others, and that this preference largely governs her response. For example, it may be that this person categorically excludes nonnative, alien, or invasive species from ethical concern. Such a preference for native species is common in conservation and restoration, and often comes out of a concern for natives and pressure that may accrue to them through competition and other apparently negative interactions with nonnatives. Since starlings as a species are not considered native to North America, perhaps she sees it as better to leave the bird to its own fate than to bring it to a rehab center where it might be treated, heal, and be released among native species. Some would refer to this bird as alien or invasive; but if its alienness is a reason for her not intervening, is the same a reason for her intervening in case one? For something to be rightly considered alien, it must be deemed to not belong in the context it is found in. If it is the case that neither the strange creature nor the starling belongs in North America, they are both alien. However, if the ecologist indeed sees both creatures as alien, it remains to be explained why she would treat each organism so differently, and in the specific ways she did. Why would she intervene in the case of the strange creature, but not intervene in the case of the "normal" creature?

The starling, it might be noted, as well as scores of generations of its ancestors, was born in North America. This brings up the question of when, if ever, a species transitions from nonnative to native status. Carthey and Banks "argue that the logical criterion for determining native status of a long-term alien species must be once its native enemies are no longer naïve" (2012: 1). Are the starling's competitors still naïve of its presence, or has it become naturalized to them? If the answer is the latter conclusion, as I suspect it is, how (if at all) would this affect treatment of the "normal" individual?

It is worth recalling that the unidentified creature was taken to a lab and not a rehab center or hospital. It would seem obvious that the reason the same was not done for the starling was because starlings are so numerous and commonly seen; because of this, much is already known about them. The strange creature, however, as far as the researcher knew, was neither numerous nor common, and could represent a new scientific discovery. It seems safe to conclude, then, that a reason for the strange creature being brought to the laboratory was its potential scientific value, and not because of any needs it was perceived to have. If this type of creature was as numerous as a starling and had no known economic or other value, it would likely be treated the same way the starling was treated. It can even be imagined that, rather than being left to its own fate, it is killed outright.

Now let us imagine a third and final scenario, in which the details leading up to the biologist's decision to intervene or not intervene are identical to the above scenarios. The biologist, however, recognizes this creature as a type of warbler, although she is unsure of the species. The possibility of it being a member of a "new" species, however, does not cross her mind. Since she believes that most warblers are native and are often rare, she carefully picks it up and transports it to the local wildlife rehabilitator.

Each of these variants on a situational theme depicts very different types of interactions and consequences for the creature in question. The first and the third scenarios include acts of intervention, whereas in the second scenario no intervention is made. For the first, the member of a potentially new-to-science (and therefore, if native to earth, likely endangered) species is moved to a laboratory. The creature's fate is uncertain, but the actions taken by the biologist reveal that, for her, the potential benefit of being involved in a scientific discovery outweighed the benefit for the organism of having quick access to proper medical treatment. The main motivation here for the biologist may well be self-interest, although it could always be argued that she was in fact motivated by contributing to the greater good of science. In either case, the reason that she was able to capture the creature at all was that it was injured and unsafe. Notice that the sense in which this creature is thought of as alien appears to have positive intervention value; let us call this alien-scarcity.

In the second scenario, no intervention is made, although the creature experienced an equivalent amount of need, injury, and vulnerability to the first. Because the group it was perceived as belonging to was not seen as threatened, it was ignored. This is despite the likelihood that, from the population-centric perspective of the biologist, helping a single starling would have virtually no effect on any of its would-be native competitor species. Unlike the first case, the alien status of this bird is perceived as having negative intervention value, although the bird acquired this status both through no fault of its own and long ago. Let us call this status alien-abundance.

In the final scenario, intervention again occurs. This time, it is to help the animal itself, and so can be considered to have occurred primarily out of interest for the other instead of self-interest. The warbler is not significant enough to warrant being taken to a lab, but it somehow deserves to be given treatment more than the starling does, even though each showed similar levels of suffering and debilitation. A single bird will hardly fortify a flagging population, so a likely explanation for the choice to intervene, in this case, is that the warbler was thought to belong here.

A relevant question, at this point, is the following: Does the difference in the types of lives experienced align with the categories deemed as relevant by the biologist? Another way of phrasing this is to ask whether similar capacities are treated as having equal worth in each of the cases. Were equals, or near-equals, treated equally due to individual need? If the correct criteria are the capacity for suffering and the will to live, then the answer is no.

It is interesting to note, however, that a highly influential figure, generally seen as a holist, does not recognize any incompatibility between ecological holism and animal welfare concerns. Arne Naess, well known as the founder of the deep ecology movement, includes the following norm in a version of his nature philosophy, referred to as Ecosophy T: "The negative value of the severe suffering of an animal belonging to a large population has a no less negative value than that of an animal of a small population!" (Naess and Mysterud 1987: 29) This norm would seem to imply that if the main goal of treatment is relief of suffering, then a suffering starling, for example, ought to be treated with the same level of care and concern as a suffering warbler.

Horta makes a case for intervention while acknowledging its potentially counterintuitive character for those focused on nature preservation. He does so by responding to some common objections to intervention and arrives at the following five points: (1) Nonhuman animals are morally considerable; (2) Animals in nature often live lives of suffering; (3) We should promote the types of intervention already occurring; (4) We are ready to implement new forms of intervention; (5) More research is warranted (2017: 24–25). While I agree with these conclusions, I would add that, in the context of wildlife rehab, sufficient justification must be present for removing the animal from its current environment.

References

Carthey, A. J., and P. B. Banks. 2012. When Does an Alien Become a Native Species? A Vulnerable Native Mammal Recognizes and Responds to Its Long-Term Alien Predator. *PloS One* 7 (2): e31804.

Horta, O. 2017. Animal Suffering in Nature: The Case for Intervention. *Environmental Ethics* 39:261–279.

Naess, A., and I. Mysterud. 1987. Philosophy of Wolf Policies I: General Principles and Preliminary Exploration of Selected Norms. *Conservation Biology* 1 (1): 22–34.

PART III

Practice

This book began with an argument and will continue with a section on practice. It is organized this way, rather than the reverse, to draw attention to the importance of thinking through the likely implications of an activity before engaging in it. While it is undoubtedly possible (and in some cases perhaps preferable) to follow a more inductive approach in theory building, beginning with observation and/or manipulation and proceeding to theory, such an approach would fly in the face of ethical conduct if generalized. What would the world be like if everyone adopted the rule "act now, think later"?

When it comes to strongly interventionist pursuits such as wildlife rescue and rehab, the more forethought given the better. The "more" I have in mind is not a temporal more, such that action is forestalled indefinitely from lack of decision or predictive power. It is a more of breadth, of assessing a wide swath of different, and likely overlapping, possibilities and perspectives. For example, although many objections to an organismal approach to conservation and care have been found lacking, it would be foolish to dismiss them out of hand before first learning about and reflecting upon alternative ethical views that would, if fitting, mandate a different set of actions.

It may be that starting with philosophizing is neither a necessary nor normal part of health-focused professions and activities. Many medical doctors, for instance, have never taken a philosophy course, and it could certainly be argued that doing this would not help with performing the services of a physician. An equivalent statement could be made about veterinarians, as well as conservation biologists. My point here, however, is not primarily about reflection aiding one's technical skills. It is that it is worthwhile to consider, if possible, the relevant range of views and approaches that characterize one's domain of interest prior to making a commitment to a subset of this domain with practical import. An effective way of doing this would likely be to liberalize, in an academic sense, scientific and professional programs. As we have seen, a certain set of ontological assumptions not unlike those inherent in an organism-centric ethical view yields perhaps the most drastic dichotomy possible: In the population-centric view, killing is permissible and often preferable, whereas in the organism-centric view, it is rarely either.

Attempts at thoroughly thinking through the implications of practice cannot be restricted to academia, however—especially given the volunteerism and broad participation within wildlife rehab. Obviously, it is unrealistic to expect every would-be rehabber to have had a thorough background in the liberal arts and sciences. Aside from being unrealistic, such a requirement would be elitist and undesirable. There seems to be no non-coercive way to ensure that would-be practitioners of wildlife rehab have considered the range of arguments and counterarguments for and against this activity prior to choosing to either engage or not engage in it. One option, which this book will hopefully augment, is for those acquainted with ethics to communicate and offer summaries of relevant questions to ponder.

It must be acknowledged, though, that the above discussion of thinking through potential consequences applies no less to those well versed in theory who wish to try their hands at rehab practice than it does to those of a more practical bent. What follows is an attempt to inform about the basics of wildlife rehab, as well as likely future opportunities for collaboration and cooperation. This section is intended to complement the education that will likely occur through training and expert guidance should one decide to try rehabbing.

CHAPTER 9

Wildlife Rehabilitation: Background

Abstract Like many historical (or perhaps more appropriately, prehistorical) activities, the precise origin of the practice of humans helping wild animals is impossible to determine. It may be epistemically risky to speculate about prehistory and the habits of prehistoric peoples, but archaeological evidence suggests that most direct human–nonhuman encounters, if non-neutral, resulted in death, injury, or domestication. To the extent that such apparently altruistic acts as are the subject of this work occurred in pre-modern times, however, they were likely isolated cases resulting from arrangements lacking a strong organizational structure. This organizational structure is still in the emergent stage for wildlife rehab.

Keywords Wildlife rehabilitation • Schweitzer • Sympathy • Preservation • Nature

Like many historical (or perhaps more appropriately, prehistorical) activities, the precise origin of the practice of humans helping wild animals is impossible to determine. It may be epistemically risky to speculate about prehistory and the habits of prehistoric peoples, but archaeological evidence suggests that most direct human–nonhuman encounters, if non-neutral, resulted in death, injury, or domestication. To the extent that

L. B. DeVaney, *Animal Conservation Ethics and the Population Problem*, The Palgrave Macmillan Animal Ethics Series,
https://doi.org/10.1007/978-3-031-91891-9_9

such apparently altruistic acts as are the subject of this work occurred in pre-modern times, however, they were likely isolated cases resulting from arrangements lacking a strong organizational structure. This organizational structure is still in the emergent stage for wildlife rehab.

Without delving into the murky depths of pre-modernity, perhaps the most likely candidate for the founder of wildlife rehab is Albert Schweitzer. Schweitzer, the 1952 Nobel Peace Prize winner, is famous for his work building and administrating a hospital in present-day Gabon (then French Equatorial Africa). Perhaps best known for his medical missionary efforts, he also took on the roles of theologian, philosopher, and organist, among others. While Schweitzer's hospital is often understood as a conventional human treatment center, it was not only used to treat humans, but nonhuman animals as well (Schweitzer 1950a). Although his practice of keeping semi-wild "pets," including a pelican (genus *Pelecanus*) and antelope, flies in the face of contemporary wildlife rehab practice, it is worth noting that the animals Schweitzer cared for, both wild and domestic, likely enjoyed more freedom than most urban wildlife today.

Schweitzer's efforts for the good of both human and nonhuman life were undergirded by an unyielding, absolutist, universalist stance on proper conduct toward living things. He referred to his position as the ethic of reverence for life. Central to this position is the belief that every living entity—regardless of size, intelligence, or even awareness—warrants ethical consideration as a direct result of its expression of the will to live. This consideration is not affected by our categories of nativity, rarity, domesticity, and the like.

Such an emphasis on the will and the desires that are considered to compose it is sometimes referred to as conativism. Callicott considers this type of position in his search for a non-anthropocentric value theory, but ultimately rejects it. This rejection is largely based on the difficulty or impossibility of following the implications of such a theory to their logical conclusion. It would seem, for example, that such a stance would prohibit even the consumption of plants (1984: 301). A conclusion this extreme, however, would only be necessary if a perfectionistic version of the view were adopted.

Schweitzer did not deny that at times hard decisions must be made, and tensions must be faced:

> Someone brought me four pelicans whose wings had been so badly slashed by insensitive people that they cannot fly. It will take two or three months

> before their wings heal and they can fly freely. I have hired a fisherman to catch the necessary fish to feed them. I always pity the poor fish to the depths of my soul, but I have to choose between killing the fish or the four pelicans who would surely starve to death. I do not know whether I am doing the right thing in deciding one way instead of the other. (Schweitzer 1992: 218)

While Schweitzer saw death, destruction, and general conflict as aspects of existence that could never be completely avoided, he nonetheless insisted upon constant mindful attention to one's actions.

The influences on Schweitzer's thought and action were many and diverse, ranging from Jesus Christ, to modern European philosophers such as Kant and Schopenhauer, to ancient Indian and Chinese thinkers. Schweitzer was, in general, harshly critical of Western philosophy's failure to give nonhuman animals adequate moral attention, as attested to by his proclamation that "Just as the housekeeper who has scrubbed out the parlor takes care that the door is kept shut so that the dog may not get in and spoil the work by the marks of his paws, so do European thinkers watch carefully that no animals run about in the fields of their ethics.... Either they leave out altogether all sympathy for animals, or they take care that it shrinks to a mere afterthought which means nothing" (Schweitzer 1950b: 297). This theoretical critique was transformed into practice through Schweitzer's move to and work in Africa.

Although Schweitzer held aspects of Asian philosophical and religious traditions in high esteem, he also disapproved of what he saw as a pervasively negative attitude toward life and being in much of the content of these traditions. This negativity in turn was seen as tending to produce passivity rather than activity for the sake of welfare. While admirable in its commitment to nonviolence, such a worldview as seen in Hinduism, for example, was considered insufficient in its failure to engender compassionate action.

When the ethic of reverence for life is compared against perhaps the archetypal ecological ethic of Schweitzer's contemporary, Aldo Leopold, some clear similarities can be seen. Schweitzer, for example, states that "The good is what preserves and advances life; evil is what hinders or destroys it" (Schweitzer 1969: 119), while Leopold asserts that something is "right when it tends to preserve the integrity, stability, and beauty of the biotic community. It is wrong when it tends otherwise" (Leopold 1949: 224–225). While both the land ethic and Schweitzer's "life ethic" are expressly about preservation, the former is more focused on the capacity for life processes than individual lives per se, thereby allowing individual welfare concerns to fall by the wayside when deemed in conflict with some perceived greater good.

Leopold is not the only link between Schweitzer and broader ecological concerns, however. Rachel Carson dedicated her landmark classic *Silent Spring*, lauded by many as the founding work of the modern environmental movement, to Schweitzer. Why Schweitzer and not his contemporary, Leopold? While Schweitzer was famous during his lifetime and has since largely faded from memory, Leopold's repute as a figure in conservation ethics has grown steadily since his death and is currently unmatched. *Silent Spring* was published in 1962, while Schweitzer was still alive, and perhaps the surrounding acclaim impressed upon Carson the importance of a popular figure to help lead the charge for nature.

Though Schweitzer's impact and reputation reached global levels, his work was local. Schweitzer's complex in Lambarene was an outstanding example of a hospital in the fullest sense: A facility that showed unparalleled hospitality. This was not only due to the structure of the facility design and layout, but also to the acceptance and treatment of patients across species lines. As Rud notes, "Hospitality here and elsewhere means being open and welcoming to the other" (Rud 2011: 93). For Schweitzer, this openness toward otherness was not restricted to needy members of the human species but was extended to any members of the animal kingdom to whom aid could be rendered.

This remarkable, trans-species concern is attested to by Charles Joy in an introductory section to *The Animal World of Albert Schweitzer*, appropriately titled "Schweitzer's animal friends." Included in this piece are some short anecdotes highlighting Schweitzer's compassionate attitude and actions toward animals. Among these accounts is Joy's following description of Schweitzer's rescue and rehab of a young antelope:

> The natives had dug a deep pit on one of the jungle trails to trap animals, and into it plunged an antelope with her small fawn. When the natives arrived, the mother in her terror gave a mighty bound and escaped. The little fawn was left behind. The natives, knowing Doctor Schweitzer's custom of liberally rewarding those who bring him for his care the helpless creatures of the forest, animals hurt in some way, or babies whose mothers have been killed, carried the tiny fawn to him. He nursed it on a bottle, kept it in a pen adjoining his study, let it roam about his table in the evening until

> it began to eat his manuscripts and nibble holes in his trousers. Every afternoon when he had leisure he took it out for a gambol around him on the river bank beneath the palm trees. He would often sit on a low wall beside his house and let Leonie lick his arm for the salt in the perspiration. (Schweitzer 1950a: 11)

The Animal World of Albert Schweitzer, which was published during Schweitzer's lifetime, was in fact dedicated to a pelican (genus *Pelecanus*): "Monsieur Le Pelican, Faithful Guardian of Albert Schweitzer." Although there is a humorous aspect to this dedication, it shows the esteem and affection toward animals held by Schweitzer. Joy writes:

> …three young pelicans had been brought to him by some natives who had captured them somehow on the sandbanks and had taken them away from their parents. Their wings were not grown, and they were weak and hungry. So the doctor had taken them under his protection, made a house for them beneath the veranda, and tried to satisfy their voracious appetites until they were strong enough to care for themselves. At first the fish had to be put into their beaks, but later they learned to catch them when thrown. Little by little they grew and acquired feathers. One of them, smaller than the others and weaker, was slow in learning to fly, but at last he too was down on the river catching his own fish.
>
> The dry season came. … It was thought that all the pelicans would leave together for the smaller streams of the interior. But the pelican who had been backward in his development stayed behind when the others went. It was evident that he had learned to like the hospital, to feel a certain security there which he did not expect to find elsewhere. There was fish to be had from the kitchen as well as the river, and there was a fellowship with other life that he seemed to like. He became fond of the African sheep. … One of the big rams became a special friend and the two were often seen together. (Schweitzer 1950a: 21–22)

Despite two of the three rescued pelicans having been successfully released, Monsieur Le Pelican chose to remain close to Schweitzer, even "guarding" his door. While such behavior may be today deemed as resulting from imprinting and therefore undesirable, the expressed autonomy from members of his own species (and apparently members of his own brood) is striking. Some of the negative effects of this semi-tame behavior, however, were experienced by the pelican in suffering a broken leg because of habitually attacking villagers to steal fish from them.

The basic guidelines Schweitzer and other early rehabbers followed are not well understood, and the extent to which such activities were structured and routinized is questionable, given their novelty and largely experimental status. To what extent did Schweitzer's human medical knowledge transfer to the nonhuman animals he was treating? How much of his practice was essentially educated guesswork, and how much was informed by previous examples? Answers to these and other questions are elusive. Specifics aside, it seems safe to conclude that Schweitzer and his assistants were early practitioners of wildlife rehab.

Today, part of Schweitzer's legacy continues in modern wildlife rehab. While a significant degree of independence continues to characterize most contemporary rehab centers, standardization of policies and procedures is increasing thanks to the growth of national and international wildlife rehab organizations and accompanying legislation. The International Wildlife Rehabilitation Council (IWRC) was formed in California in 1972 "by a group of individuals concerned about the care and conservation of native wildlife" (IWRC History 2024). The IWRC and its members now publish the *Journal of Wildlife Rehabilitation*, host a yearly symposium, develop and lead courses and training for becoming a Certified Wildlife Rehabilitator (CWR), and offer handbooks and other resources.

A decade after the founding of the IWRC, the National Wildlife Rehabilitation Association (NWRA) was formed following the first National Wildlife Rehabilitation Symposium in 1982, hosted by Willowbrook Wildlife Haven and the Forest Preserve District of DuPage County in Naperville, Illinois (NWRA About Us 2024). Hardcover copies of the proceedings from this first NWRA symposium were published in the volume *Wildlife Rehabilitation*, spanning a wide range of topics from philosophy, history, and advancement of individual rehab programs and centers, to highly specialized medical procedures such as construction of artificial legs for birds (Beaver 1982). The NWRA publishes the journals *Wildlife Rehabilitator* and *Wildlife Rehabilitation Bulletin*, as well as many guidelines and resources such as the *Wildlife Rehabilitators Code of Ethics* and the *Wildlife Educators Code of Ethics*. In 2021, the IWRC and NWRA collaborated to publish *Standards for Wildlife Rehabilitation*, based on previous editions of *Minimum Standards for Wildlife Rehabilitation*. The NWRA estimates that over 75% of admits are "affected in some manner by human activities" (NWRA About Us).

Despite the communication and cohesion facilitated by such groups as the IWRC and NWRA, rehab rules and regulations differ according to location and type and rarity of animal. Rehab of most birds requires the issuance of a federal permit by the US Fish and Wildlife Service, due to the US Migratory Bird Treaty Act; the only exemptions are members of non-native species. State laws vary greatly, and in Minnesota are handled by the state DNR. Three permit classes exist in Minnesota: Novice, general, and master. Novice rehabbers may only admit orphaned (not sick or injured) birds and small mammals; treatment of endangered, threatened, or raptor species is not allowed. A potential general rehabilitator must have held a novice permit for at least two years prior to applying. In addition to novice privileges, a general may possess and treat sick and injured of the same species allowed for novices, as well as some small raptors, deer, and bears if specified in the permit. A potential master rehabilitator must have held a general permit for at least four years prior to applying. A master may possess all the foregoing, as well as big game and endangered and threatened species, if specified in the permit and in accordance with federal permitting rules. Applicants may attempt to obtain a permit of a desired class by passing a written exam and a facility inspection. The Minnesota DNR has a Wildlife Rehab Advisory Committee (WRAC) for master rehabbers who wish to participate. Permit holders of all classes must have a licensed veterinary adviser with whom to confer (MNDNR Wildlife Rehabilitation).

It is widely acknowledged today that practices under the broad rubric of directly helping wild animals can be placed in four categories: Rescue, rehab, release, and research. These "four Rs" form the core of what is broadly known as rehab practice, with rehab itself the central piece (Pyke and Szabo 2018). Before the first "R" (rescue) can be attempted, however, one must make a judgment as to whether the animal being considered truly needs help, i.e., whether it is, or is in imminent danger of, facing an emergency. Although the cues to need of aid will differ from taxon to taxon and species to species, Hentz lists the following as universal signs of an emergency for an adult animal: (1) bleeding; (2) unconsciousness; (3) abnormal use or position of limb(s); (4) odd head position; (5) maggots; (6) was in a pet's mouth; (7) drainage from eyes or nose; (8) behaving abnormally; (9) cold and lethargic; (10) heat exhaustion and exposure; and (11) inability to escape (2009: 38). Although helpful as guidelines, most of these signs are not unambiguous and therefore require a judgment call on the part of the would-be rescuer, as well as some background knowledge of the normal behavior of members of the affected species.

Release and research go hand-in-glove and will likely be increasingly paired as rehab outcomes and their relevance to conservation are recognized as important. In addition, rescue databases have been identified as a crucial piece of future research (Pyke and Szabo 2018: 56). Currently, funding is often a barrier to performing extensive rehab-related research. However, adoption of citizen scientist, volunteer-based approaches and student-led initiatives is possible when a lack of financial support exists. While the acknowledgment of research as a critical component of the rehab-centered process is relatively new, it bodes well for the success of future efforts.

References

Beaver, P., ed. 1982. *Wildlife Rehabilitation: Volume 1.* Smithtown: Exposition Press.

Callicott, J. B. 1984. Non-Anthropocentric Value Theory and Environmental Ethics. *American Philosophical Quarterly* 21 (4): 299–309.

Carson, R. 1962. *Silent Spring.* Boston: Houghton Mifflin.

Hentz, P. S. 2009. *Rescuing Wildlife: A Guide to Helping Injured and Orphaned Animals.* Mechanicsburg: Stackpole.

International Wildlife Rehabilitation Council. 2024. *History.* History—International Wildlife Rehabilitation Council. https://theiwrc.org/history/. Accessed 11 December 2024.

Leopold, A. 1949. *A Sand County Almanac and Sketches Here and There.* New York: Oxford University Press.

National Wildlife Rehabilitators Association. 2024. *About Us.* National Wildlife Rehabilitators Association. https://www.nwrawildlife.org/page/AboutNWRA. Accessed 11 December 2024.

National Wildlife Rehabilitators Association and International Wildlife Rehabilitation Council. 2021. *Standards for Wildlife Rehabilitation.*

Pyke, G. H., and J. K. Szabo. 2018. Conservation and the 4 Rs, Which are Rescue, Rehabilitation, Release, and Research. *Conservation Biology* 32 (1): 50–59.

Rud, A. G. 2011. *Albert Schweitzer's Legacy for Education: Reverence for Life.* New York: Palgrave Macmillan.

Schweitzer, A. 1950a. In *The Animal World of Albert Schweitzer: Jungle Insights into Reverence for Life*, ed. C. R. Joy. Boston: Beacon Press.

Schweitzer, A. 1950b. *The Philosophy of Civilization.* Trans. C. T. Campion. New York: Macmillan.

Schweitzer, A. 1969. *Reverence for Life.* Trans. R. H. Fuller. New York: Harper & Row.

Schweitzer, A. 1992. *Letters 1905–1965.* Trans. J. Neugroschel. New York: Macmillan.

CHAPTER 10

Minnesota Rehab: A Tale of Two Centers

Abstract Wildlife rehab in Minnesota, as in many states, occurs in both home-based and institutional settings. What follows are brief recaps of insights gleaned from in-depth interviews with two leading figures in Minnesota rehabbing. The first recap is derived from a recorded conversation with Peggy Naylor, former director of Wildwoods Rehabilitation in Duluth. Wildwoods is a general class facility and does not accept threatened or endangered species. The second recap is derived from a recorded conversation with Phil Jenni, director of special projects for the Wildlife Rehabilitation Center of Minnesota (WRC) in Roseville. The WRC is a master class facility and accepts threatened and endangered species.

Keywords Minnesota • Interview • Wildlife rehab • Education • Wildlife health

Wildlife rehab in Minnesota, as in many states, occurs in both home-based and institutional settings. What follows are brief recaps of insights gleaned from in-depth interviews with two leading figures in Minnesota rehabbing. The first recap is derived from a recorded conversation with Peggy Naylor, former director of Wildwoods Rehabilitation in Duluth.

L. B. DeVaney, *Animal Conservation Ethics and the Population Problem*, The Palgrave Macmillan Animal Ethics Series,
https://doi.org/10.1007/978-3-031-91891-9_10

Wildwoods is a general class facility and does not accept threatened or endangered species. The second recap is derived from a recorded conversation with Phil Jenni, director of special projects for the Wildlife Rehabilitation Center of Minnesota (WRC) in Roseville. The WRC is a master class facility and accepts threatened and endangered species.

Wildwoods

I began the interview by asking Peggy how she got involved in wildlife rehab. She recalled her sister-in-law finding a wounded pigeon in Austria and bringing it to a vet, who said she would bring it to a wildlife rehabilitator. This was the first time Peggy had heard of wildlife rehab.

At this early stage, the interview was broken by a gray fox sighting in Peggy's backyard, which contained many of Wildwoods' enclosures. This was the first time Peggy had seen a fox in their yard, and the first time I had seen a gray fox in the "wild." The encounter left me with a sense of newness associated with the project.

After the fox surprise, Peggy discussed the origins of Wildwoods. She and her husband initially trained under Gail Buhl, then at Wolf Ridge Environmental Learning Center, now Education Program Manager at the Raptor Center. She mentioned starting out as a 501c3 nonprofit in 2010, prior to which donated funds were transferred to the Raptor Center and WRC.

The second interview break occurred at this point, as a volunteer brought over a house sparrow that had died en route to Wildwoods. The discussion then turned to an idea for a new facility for Wildwoods within an ecological consortium, which might be created on the site of an old school with 17 acres of land for sale. In addition to a rehab center, this consortium could include a folk school, senior assisted living, rentable space for nonprofits, and community gardens—possibly including gardening for wildlife. At the time of the interview, plans had already been drawn up for a 5000 square foot center.

Considering Peggy's comment that education is at least 50% of the mission of Wildwoods, I bring up the issue of educational licensing. Peggy explains that Wildwoods does not have education animals but uses Facebook and other social media as outreach tools. Peggy refers to the dead sparrow just brought to her, which was killed by a domestic cat. She mentions that a photo of the bird is going on Facebook, along with statistics regarding cat predation.

The third interview break occurs as Peggy sees a northern goshawk. Peggy gives the following analysis of the situation: The goshawk is interested in the squirrels, and sometimes eats pigeons and robins being rehabbed under a soft release routine. That is just what it's like to be a wild animal. However, attacks from unrestrained pets are completely different and inexcusable.

I mention I had not heard of soft release before, and the distinction between soft and hard release is explained. Hard release occurs when animals are immediately let go to fend for themselves, with no support, whereas soft release occurs when food is provided for transitional aid. An example of soft release is given: Whistling in robins to a box of dirt and worms at feeding times.

The problem of getting "squirreled out" (when an area is overpopulated by squirrels) is referred to. Other Minnesota rehabbers are discussed, along with a cooperative arrangement between them and Wildwoods. Peggy recalls only two other (individual) rehabbers in northeastern Minnesota. This lack of rehabbers brings up the problem of transportation of patients to other facilities when necessary; however, tapping into traffic between the Twin Ports and the Twin Cities seems to work. Around 450 patients were admitted to Wildwoods in 2011; by 2015, the total had surpassed 1000.

Peggy brings up the problem of licensed volunteers getting overwhelmed when filling in, and recently sending about 20 squirrels to WRC in Roseville. This would not have happened if Peggy and her husband had been there. Squirrels and bunnies are Wildwoods' focus, and most baby birds get sent to WRC or the Raptor Center. I asked if Wildwoods had good help and was told that about one in 10 trainees end up being reliable. Peggy and her husband do at least half of the feeding shifts themselves—some days they do them all.

I ask about Wildwoods' relationship to the University of Minnesota-Duluth (UMD) and am told that the father of the volunteer who had brought in the dead sparrow is the head of the animal lab. UMD supplies cages, and Wildwoods sometimes donates specimens—they have a collection permit—but that is the current extent of the relationship. I inquire as to whether Peggy would want Wildwoods to have a closer affiliation with UMD in the future, and the response is largely negative. She wants to be self-controlled and does not wish to partner with any but trusted rehab-related organizations. She is in favor of Phil Jenni's model of the WRC being the hub and Wildwoods being a spoke.

I ask how many vets Wildwoods has on staff and am told that (unlike the WRC) there are none, but around six that are consulted on a volunteer basis. Most of them do not specialize in wildlife. I ask if Wildwoods gets any government funding, and Peggy says no. Funding mostly comes from individuals, although they recently got some funds from a community center and the local Biodiversity Fund for educational programming. Charismatic species (e.g., bobcats) tend to bring in money.

Other fundraising options we discuss include an adopt-an-animal program. Wild and Free already does this—for $100 you can sponsor an animal and get a picture and a story, and sponsors can be present at the release if it's not a high-stress animal like a bunny. I ask if rehab is useful for conservation, and I am told that it is only insofar as it gets people hooked into the wider issue. Peggy expands by referencing the ripple effect and the importance of taking advantage of teachable moments—for instance, when an animal is brought in.

I ask whether Peggy comes across any hard problems concerning the good of animals versus ecosystems, and she denies that this is particularly problematic. The example of pigeon egg removal in Bern, Switzerland, is given as a success story in this regard. Lofts and feeding stations have been built around the city, and volunteers collect and destroy the eggs twice per week, which can cut pigeon numbers in half in a year. Peggy is considering proposing this model for Duluth. The danger of species transfers is brought up (e.g., starlings and house sparrows being released in Central Park in New York). Although they should not be here, it is not their fault—implying that we have a duty to care for them and/or look out for their interests.

I ask Peggy what she would like to see change in wildlife rehab in the future. She quickly cites the need for the acceptance of skunk rehab. Skunks have the same rabies risk as bats and raccoons, which can currently be rehabbed. A story is told of someone shooting a mother skunk, and then asking what to do about babies in the yard. All that can be done is putting out watermelon and cat food and hoping for the best. The benefit of skunks aerating the yard is discussed, as is the issue of caring about babies but not adults.

The challenge of being "all things to all animals" is brought up, as is the luck of having three centers of excellence within a two-and-a-half-hour drive. The goodness of Phil Jenni's model is reaffirmed. Although

Minnesota has many unmet rehab needs, it is strong compared with North and South Dakota, neither of which has any rehab.[1]

Phil Jenni is lauded as a visionary. Peggy states that, if the WRC had not been there and so supportive, they likely would have thrown up their hands in despair five years ago. I ask what happens when advanced medical care is needed, and I am told that if it is fancy, it must go down (to the Twin Cities). All vet work done for Wildwoods is pro bono. In closing, I ask if Peggy has a favorite kind of animal to work with, and she says that whatever animal she happens to be treating is her favorite.

The Wildlife Rehabilitation Center of Minnesota

Minnesota's "flagship" facility, the Wildlife Rehabilitation Center of Minnesota (WRC), began in 1979 as a student club of the University of Minnesota Veterinary College. Initially housed in the Old Veterinary Anatomy Building[2] (originally the Veterinary Medicine Building) on the St Paul Campus, the WRC moved to a new, state of the art facility in Roseville, Minnesota, in 2003. This is the location where I engaged in participant observation of wildlife rehab.

According to Phil, the need for regional centers (within the state) is real, since not all areas of the state have easy access to a rehabber, which can hinder a needy animal from receiving any care or treatment. This brings up discussion of a possible statewide wildlife health system, which would be analogous to human health systems. For example, the current Roseville location is basically a trauma center (like Hennepin County Medical Center), and other specialty centers could be developed. WRC is trying to get nurseries down to the Inver Grove Heights campus (outdoor caging was built there in 2008), since there is far too much activity in Roseville for a healthy animal. Another possibility is to partner with humane societies. In 2008, Phil brought up to Peggy the option of Wildwoods as a WRC "satellite" location in Duluth.

[1] Through subsequent conversations with a faculty member at South Dakota State University, I learned that this belief is not true. Some rehab does presently occur in South Dakota, while rehab previously occurred in North Dakota, which has since ceased (specific time frames and reasons were not given). According to North Dakota Game and Fish, treating injured wildlife is illegal in North Dakota (ND Game and Fish Orphaned and Injured Animals 2024). However, Minot zoo veterinarian Logan Wood became North Dakota's first certified wildlife rehabilitator in 2021 (*Minot Daily News*).

[2] This building was demolished in 2011.

I asked Phil how he initially got involved in wildlife rehab. Phil was the chief operating officer at the Citizen's League prior to the WRC. A friend of his who is a nonprofit turnaround expert became short-term executive director at the WRC. She emailed him about the job opening, and he thought it looked interesting but didn't apply. However, he saw her later at an event and asked her how the search went. She hadn't yet found a good candidate, but offered him an interview, and he got the job in March of 2002.

We discussed the origin of the WRC as a veterinary student club in 1979, with the Raptor Center leading the way a year or two before. The WRC acquired 501c3 nonprofit status in 1996 and moved off the Saint Paul campus in 2003. The old building was falling down and full of cockroaches, and the center had no professional attitude at that time. Still, they have a stronger connection with the university now than they did before. Phil is on the Dean's Advisory Committee and has received an Outstanding Service Award from the College of Veterinary Medicine.

The WRC works with individual permit holders, as well as smaller centers such as Wild and Free in Garrison.[3] For years the WRC would transfer all deer to a woman in the Saint Cloud area, Linda Peck, who had a soft release program. However, the DNR changed rules and now deer must be released back into the county where they were found because of fear of chronic wasting disease. They also work with Connie Lafond in Maple Plain for foxes, and Wildlife Rescue and Release (WRR), which is sort of a trade group of individual rehabbers. WRR works with the Animal Humane Society (AHS) in Golden Valley. The WRC also has the Orphan Outreach Program (OOP), which is a network of volunteers who care for very young orphans. While patients can be brought to the AHS in Golden Valley, they are then sent to individual rehabbers who become responsible for everything, including paperwork. With the OOP, the patients are brought back to the WRC prior to release.

The WRC had 8914 admits in 2012; by 2021, this number had more than doubled to over 19,000 (Jenni 2021). The AHS wants to work with the WRC, but the WRC is wary to take on the extra load. Sandhill cranes are usually sent to the Raptor Education Group in Antigo, Wisconsin, which takes more than just raptors, and has non-releasable cranes which act as foster parents for orphans.

[3] This and all other place names in this paragraph refer to communities in Minnesota.

The WRC has roughly 600 yearly volunteers, with a core group of about 140. Some of the most loyal volunteers have been with them for 15 years. Also, interns come from all over the world, although they work without pay for 30–35 hours per week. The WRC accrues roughly 43,000 yearly volunteer hours and has about 35 people on year-round staff. For funding, they take tax-deductible contributions. Most contributions come from individuals, because not many corporations have wildlife in their guidelines.

Around 12,000 people give the WRC money in an average three-year period. Roughly 85% of the WRC's $2.8 million annual budget comes from these contributions. The highest contributing donor gives $65,000 per year. Phil refers to the profile of donors as a demographer's nightmare. Phil (at the time of the interview) is 58 years of age, and he says that a statewide rehab system will not be done by the time he retires. He brings up plans for a main satellite facility on 65 acres in Inver Grove Heights, which someone he knows (Vance) wants to donate. They already have outdoor caging down there; it's the last stage before release for many birds, both avian and waterfowl. The WRC has also purchased 22 acres in Grant, where plans to build an additional facility are being made (Gustafson 2024: 27).

According to Phil, Vance must convince his neighbors to put land into conservation easements, although a neighbor to the north already has cul-de-sac housing projects planned out. He has been getting money from the Dakota County Open Space Program, and Dakota County has been getting money from the Legacy Amendment (sales tax money). Phil discusses further the idea for Vance's house as administrative and maintenance headquarters. It may also serve as housing for interns. I bring up that what I'm hearing sounds like the beginning workings of a rough hub and spoke model, and Phil agrees.

The possibility of the DNR outsourcing permitting duties, including on-site inspections, is brought up. This would be new, as in most other states state agencies handle permitting. However, this would likely be welcomed by the DNR, as wildlife rehab is ranked low on their priority list.

The WRC is helping build the most advanced database in the country for tracking wildlife disease with the United States Geological Survey in Madison, Wisconsin. This could become the equivalent of the Center for Disease Control for wildlife. There is an increasingly acknowledged link between human and animal disease, which is addressed through the University of Minnesota's School of Public Health, the Minnesota Animal Board, and programs like One Globe One Health. This is framed as a relationship between systems.

Rehab, in Phil's view, is insignificant for the population level. There are clearly linkages, however, between schools of agriculture, veterinary medicine, and public health. Despite lack of population effects, Phil points out that rehab makes a big difference for individuals. It is a piece of a compassion ethic building into a conservation ethic. However, this view is not shared by everyone in environmental fields, as evidenced by hostility Phil, as a Ramsey County Park Commissioner, received from a Ramsey County Naturalist (who is now friendlier to rehab in general).

Phil thinks that, other views notwithstanding, generalist nonnatives must be seen as evolutionarily impressive. According to Phil, "the bottom line is to try to create a real understanding of what motivates them and what motivates us and look to the best in both of those worlds and try to blend them together." Who the "them" and "us" are isn't plainly clear, but I assume ecologists/conservationists and rehabbers, respectively, are being referred to.

I ask Phil what he would like to see change in rehab in the future, and he replies that the professionalism of rehab needs to increase. In Phil's view, as some extreme environmentalists are bad for nature, so some extreme rehabbers are bad for the business of rehab. How, exactly, this professionalism should increase is left open for the time being.

References

Gustafson, A. C. 2024. Creature Care. *Minnesota Conservation Volunteer* 87 (516): 20–31.

Jenni, P. M. 2021. *Wildlife Rehabilitation Center of Minnesota 2021 Annual Summary.* https://wrcmn.org/content/348/00665f61/2021-WRC-report-.pdf. Accessed 12 Jun. 2025.

Minot Daily News. 2021. Minot Zoo Veterinarian ND's First Certified Wildlife Rehabilitator | News, Sports, Jobs—*Minot Daily News.* https://www.minotdailynews.com/life/arts-entertainment/2021/06/minot-zoo-veterinarian-nds-first-certified-wildlife-rehabilitator/. Accessed 11 December 2024.

North Dakota Game and Fish Department. 2024. Orphaned and Injured Animals | North Dakota Game and Fish. https://gf.nd.gov/wildlife/orphaned. Accessed 11 December 2024.

CHAPTER 11

Participant Observation and Reflections on Rehabbing

Abstract This study was conducted using the broad research method of participant observation (PO), the written results of which are often referred to as ethnographies. Franz Boas, known to many as the father of American anthropology, has been credited with founding the method (Stocking 1960). Jorgensen defines PO in terms of seven core features:

Keywords Participant observation • Insider • Volunteer • Social science • Ethnography

Methodology

This study was conducted using the broad research method of participant observation (PO), the written results of which are often referred to as ethnographies. Franz Boas, known to many as the father of American anthropology, has been credited with founding the method (Stocking 1960). Jorgensen defines PO in terms of seven core features:

1. A special interest in human meaning and interaction as viewed from the perspective of people who are insiders or members of particular situations and settings;

L. B. DeVaney, *Animal Conservation Ethics and the Population Problem*, The Palgrave Macmillan Animal Ethics Series,
https://doi.org/10.1007/978-3-031-91891-9_11

2. Location in the here and now of everyday life situations and settings as the foundation of inquiry and method;
3. A form of theory and theorizing stressing interpretation and understanding of human existence;
4. A logic and process of inquiry that is open-ended, flexible, opportunistic, and requires constant redefinition of what is problematic, based on facts gathered in concrete settings of human existence;
5. An in-depth, qualitative, case study approach and design;
6. The performance of a participant role or roles that involves establishing and maintaining relationships with natives in the field; and
7. The use of direct observation along with other methods of gathering information. (1989: 13)

I will now briefly describe how the research I performed aligns with each of these seven features, as well as a few noted points of departure. First, I sought to develop the viewpoint of an insider through taking on the role of volunteer (and later, crew leader) at the WRC. This role was learned through participation in and habituation to myriad everyday tasks unique to the setting. Given the ready-made volunteer role for this setting, formal insider status was rather quickly and easily gained, but not automatic. The steps I took to obtain this status were discovery of the WRC through finding an injured wild animal and asking for advice from a bystander about what to do with it; bringing the animal to the facility, where it was admitted; signing up to receive emails from the WRC at this visit; receiving news of an open house via email, attending the open house, and learning of volunteer opportunities; attending a short volunteer orientation and training session; and engaging in my first work shift. Thanks to the strong organizational structure of the WRC, these first steps were not difficult to identify and take. Gaining trust and respect, though, took a great deal of time, energy, and dedication.

Progress was made from nearly pure observation at the outset, to a high degree of participatory immersion. While I was performing most volunteer tasks by the end of the first two weeks in each nursery (except for the waterfowl nursery), it took months for them to become routine and habitual. A sustained attempt was made to remain observationally astute for the time span of the project, but this became difficult to the degree that efficiency was a focus. During this roughly two-year work and study period, research was conducted on various aspects of wildlife rehab, to become more familiar with the field in general and gain knowledge of typical practices and research issues.

While I certainly took interest in the human meanings ascribed to the field and activities performed, as is evidenced in sections of my field journal as well as the interviews conducted, this was not my main concern for the study (for a motivational and affective account of rehabbing, see Kidd et al. 1996). More to my design is an account of the unique setting and process of institutional rehab, from the perspective of a particular insider, and the sorts of actions and interactions afforded by the situations inherent to this setting. Since the study is based on the viewpoint of a single researcher, it cannot be comprehensive; but hopefully what it lacks in breadth it makes up for in depth. Moreover, I took as my focus not only actions strictly within the human sphere, but also interactions between humans and other animals. This approach is part of a broader call in sociology to include nonhuman animals within the purview of sociological (and other social science) inquiry (Peggs 2012).

Second, the context for this study is a new setting in which life-preserving activities are currently carried out at a small scale. Wildlife hospitals are special in their status as organizations that are emergent yet provide one of the most basic services available to living beings: The maintenance of life. Although questions about the future prevalence of similar facilities, and the use of the WRC as a model or framework for such potential facilities, are relevant, such possibilities can only be actualized through a more thorough understanding and dissemination of information about the structure and function of those that currently exist. Comparative studies of facilities are needed, but because of the newness and scarcity of wildlife hospitals even in developed nations, basic research of individual, exemplary facilities is first in order. The need for descriptions of good models is immense, as there are still states with no rehabbers at all—let alone a single state-of-the-art facility (P. Naylor, pers. comm.).

Because of the pressing need for quality wildlife rehab, inquiry about and description of what goes on in concrete settings is much needed not only for generating awareness and interest, but also as a fleshed-out example for those who might be interested in rehabbing but are concerned with the apparent lack of exemplarity in the field. While federal and state policies on rehab are readily available, and practitioners' manuals, guidebooks, and the like increasingly so, there are few thick descriptions of actual settings where rehab is performed. The exploratory nature of PO, while causing it to appear less than rigorous to some, is uniquely suited to generating a large amount of a variety of types of useful information within a moderate time frame.

Third, part and parcel of the rationale for PO as a method is its flexibility, and the necessity of such flexibility for addressing varying and unpredictable circumstances. More rigid research approaches such as traditional experiments and surveys, while appropriate for the study of highly controlled phenomena, are ill suited to detecting the interactional subtleties that tend to emerge out of naturalistic settings. Humans and their interactions with each other are notoriously difficult to predict (Ross and Nisbett 1991), and this lack of predictability is increased in our case through both technological mediation and the fact that wildlife rehab is a practice that is both with and for other animated species. What we have here are the ingredients of a very novel situation or set of situations.

Another way of making the above point is to say that, compared with PO, such more conventional scientific methods are phenomenologically thin. This thinness becomes especially evident when the setting being studied is itself not well known to outsiders, but potentially relevant to them. Science begins with observation, and control only becomes possible (but not necessarily ethically appropriate) once a phenomenon is well understood. Moreover, it is the case with wildlife rehab that the use of a limited, temporary form of control is sometimes necessary to accomplish the ends of freedom and flourishing.

While I acknowledge the importance of PO for crafting theory, I think it is helpful to make a distinction between two levels of theorizing that are occurring in this project. There is abstract ethical theorizing, writ large, about rehab as an activity that intervenes in the lives of wild animals for their benefit, mixed with metaphysical theorizing that attempts to answer the question of what constitutes conservation biology. However, there is also a more concrete level of theorizing that is distinct from the first, primarily in virtue of its descriptiveness. Therefore, it must be readable, precise, and rather open-ended. This latter type is the kind of theorizing that PO is good at helping promote. In addition, it may be helpful to point out that the former is more properly theory by itself, while for the latter the finer points of theorizing are left up to the person doing the reading and interpreting. So, while the theory being presented in this work is by no means entirely derived from PO, neither is it purely speculative and without grounding in concrete reality.

Fourth, the basis for the latter, finer-grained theorizing is the context and setting within which the study was conducted. Although speculation is always possible when applied to vague or abstract notions, the empirically grounded theory produced by PO could not have been given at the outset, but its emergent nature required practicing patience and diligence in capturing details as they presented themselves (Bryant and Charmaz 2007). This research was about observing and participating in a process of inquiry that, because of its very nature, could not render immediately valid results. The questions asked did not even stay static, but gradually conformed themselves more and more to the research setting. In a sense, the questions asked appeared to come to gradually correspond with, i.e., become more like, the answers.

This conforming of questions took place through descriptive accumulation and resulted in the net effect of increased specificity. For example, while an initial question prior to starting volunteer work was "How are the nurseries arranged?" after sustained interaction and time spent in the various nurseries, the question broke up into related but more detailed questions, often comparative, about the specific nurseries, such as "Why does the avian nursery have several incubators while the mammal nursery has none?" This increase in specificity helps explain how one can seem to know less about something after sustained study than before one had any experience of it. Although one does not literally know less at this point, the fact that one's question "tree" has so many more branches now than before gives the illusion of a massive and acquired ignorance.

Thanks to its human-animal-technological nature, the setting of most of the study is so complex that the relevant occurrences could not have been immediately predicted with any degree of accuracy. Because of this complexity, the mode of inquiry was different not only from the standard scientific method, but also the philosophical method. When a question is formed about the domain of thought or idea, as in philosophy, there is no need to wait and see what happens. It may be the case that the answer does not come quickly, or it may be the case that it does. Likewise, it is becoming more conventional in science to not perform experiments but make predictions based on simulations. Even if an experiment is performed, every possible detail of the setting is controlled in such a way that manipulation can be used to speed along the process to the desired result. In a naturalistic setting, however, one may not in such a way vex or torture nature, as readers of Bacon have variously put it (Pesic 1999), to obtain fast-track results via experimentation or modeling. This is partly because it would be impossible to control every interaction of every variable even if one wanted to—especially if the identity of rehab as a healing practice is to be retained. It is also, however, because the "whatever else happens" is not seen as just noise or distraction in PO, but the main event(s).

Whether or not Bacon intended to convey the specific notion of torture in relation to experimentation, this analogy is interesting. While rehabbing is not experimental in the scientific sense of the term, it is experimental in the broader sense of being largely trial and error. Moreover, while most rehabbing procedures are probably not truly torturous, some are clearly not enjoyed by the patients (such as being injected with fluids). A disanalogy lies in the fact that the procedures are performed because they are thought to be to the patients' physical benefit, whereas torture—whether performed in the name of science, politics, religion, or whatever—is normally either for the benefit of the torturer and/or associates, or for the purpose of ritual purification or spiritual cleansing. Rehabbing, then, could be described as unconventionally both experimental and torturous, but not conventionally so. The opportunism inherent to PO, while frowned upon by some academics, is essential for studying phenomena that are largely unknown to outsiders. A flexible design framework is helpful for taking advantage of the various data-gathering opportunities that will present themselves, some unexpectedly. Examples of these unexpected opportunities included the many anecdotes derived from seemingly random wildlife-oriented encounters I experienced during the study. Had I not been open to the possibility of such scenarios, but categorically excluded them at the outset because of their lack of predictability or regularity, I would have been deprived of key information.

Fifth, the approach taken is strongly qualitative in character. While numbers are used at times, mostly to show the sheer quantity of animals being cared for at the WRC, these quantities are not the central feature of the research. One of the main reasons I have tried to keep the data qualitative is because of the conscious effort to build a database that is readily understandable and interpretable by humans, in the narrative analysis tradition (Daiute and Lightfoot 2004). While numbers give the appearance of precision and scientific rigor, they cannot capture the range of concepts, attitudes, beliefs, values, and other aspects of human life, nor can they adequately convey them. This study was not about compressing lived realities into bits and bytes, but about uncovering, from a human viewpoint, what it is like to help other animals in a specific organized setting. Because of this specificity and detail, the study is also case-based and in-depth. Interviews of key figures were conducted, and the option of conducting broader social science research, e.g., about the characteristics of rehabbers in general, was eschewed. While an underlying hope is that the model of the WRC will be generalizable to other areas, this was ancillary to the goal of performing much needed basic research on a unique facility and setting.

Sixth, although compared with many potential and actual PO settings the discussed participant role was readily available, it was challenging at times to establish and maintain rapport. A difficulty perhaps unique to volunteer settings is that, while it is relatively easy to gain entry to the setting in question, it is often quite difficult to gain more than a minimal amount of respect from those in authority. This, I believe, can be due to two factors: One, it is known by those coordinating volunteers that said volunteers likely did not earn their position competitively; and two, in cases where the volunteer coordinator occupies a paid position, they may fear for their own job security if especially adept volunteers join the crew. The latter was not the case at the WRC, since crew leaders are generally not paid, but, in my experience, the former was a factor. The challenge of this situation is that the crew leader is, in the name of fairness and caution, forced into the rather awkward position of assuming that each volunteer under their supervision is essentially unskilled. This assumption is especially important in fields such as wildlife nursing, where lives are obviously at stake. The awkwardness is only compounded, moreover, in the case that, after time with volunteers and their habits, favoritism begins to emerge.

In my experience, a third stage of difficulty was reached when this favoritism, preference, or apparent judgment of quality led to the offering of crew leader roles to volunteers for the same crew they had worked with since the beginning of the season. At this point, it became less a matter of maintaining relationships with "natives" (however unfitting this designation may be) in the field, and more a matter of managing relationships with those whose statuses have now been lowered compared with one's own. At the point I was converted from plain volunteer to crew leader, I sensed some negativity on the part of my crew members. It was not animosity or contempt, but more a questioning of why they were not chosen. I wondered this, as at least one of them I believed to be more competent than I.

An interesting aspect of conducting PO in an "out" or visible organization setting is the relative immunity one has from being expelled from the context of the setting. In an invisible setting such as the drug dealing scene (Hindmarch 1972), for example, developing and maintaining relationships is central to study success; in such a context, the moment you lose rapport, you are no longer an insider, nor likely a participant observer. This is less the case in organizations visible to the public like the WRC, where the transparency and organizational hierarchy serve to insulate from such outcomes. While I have heard crew leaders voice negative perceptions of volunteers, I have yet to encounter an attempt to exclude a volunteer from the privileges of insider status.

Seventh, while direct observation and participation were the main means of gathering data for the study, other methods were used as well. In-depth interviews were conducted with key figures in the local rehab scene, and theorization derived from philosophical knowledge and texts, as well as more conventional scientific findings, was worked into the main argument for focusing conservation at a smaller ecological scale than has been the norm. Moreover, observation and participation were not restricted to the setting of focus but were extended to a broader area. Relevant accounts from different states, as far away as the East Coast of the United States, were recorded.

Although my project includes aspects of each of these seven features, it is unique in that it is focused on a practice and organizational setting that is marked by concern for nonhumans and nonhuman experiences. The most noteworthy departure of this study from the norm, then, is its distance from the expressly humanistic stance taken by both the author of *Participant Observation* (Jorgensen 1989) and most PO studies to date. Rather than being classified as purely anthropological or sociological, this project should be seen through the broader, more inclusive lens of anthrozoology—the study of human–nonhuman animal interactions, also known as human-animal studies. Moreover, not only is it about animals, it is about a practice that is strongly for animals. This ethical aspect marks another clear point of departure for my project from the PO norm. Wildlife rehab, while certainly interesting, is an activity practiced not merely out of interest, but out of values and a sense of duty.

A common way of characterizing PO studies is according to the relative level of observation and participation performed, on a scale from nearly purely observational to almost entirely participatory (Ibid.). Because of the strong ethical component of rehab noted above, my role, while observational, was very strongly participatory. As critical as it is to soak up, think about, and record every observation and experience possible while in the field, when wildlife nurses are in short supply (as was often the case) the first concern is the effective and efficient performance of one's assigned duties.

Another contrast often posed in relation to PO is that of the insider versus the outsider. As a researcher gains experience with a setting and the varied events and situations common to it, he or she is said to gradually move from the position of outsider to that of an epistemologically and socially privileged insider. I found these measures of insider privilege to be asynchronously developed in my case. While the strong institutional and organizational context of the WRC along with the volunteer opportunities offered allowed me nearly instant social privilege, knowledge, much of which was experientially derived, was gained slowly and incrementally.

I will now move from a description of the methodology of PO to an account of themes derived from my experiences at the WRC. These themes are derived from two years of PO in the avian, mammal, and waterfowl nurseries. While they are by no means comprehensive, the hope is that they can help orient interested individuals to the context of rehabbing. In addition, the below account can help elucidate what the process of using PO is like in practice.

Themes from Participant Observation

One of the basic, defining themes of my time at the WRC was emotion. This can be understood as a singular category broadly separated from reason, as well as broken into myriad distinct types. Several emotions were felt during the rehabbing process, some of which are described below. While the emotional content of every experience was not always immediately apparent, upon reflection the underlying feelings often became clearer. At first, the flood of sensations was difficult to deal with: "My first experience volunteering at the WRC was characterized by feelings and emotions of stress, anxiety, hurriedness, and helplessness, along with joy, gratitude, excitement, and even amusement. The more negative of these emotions, I surmise, resulted from my complete lack of experience in the field of wildlife rehab, which was greeted by a barrage of new sensory data upon entering the Avian Nursery (my assigned volunteering area)."

Communication was a key to rehabbing success. Effectively communicating a wide range of information was crucial as I progressed from observer to crew leader. Eventually, human-to-human nonverbal cues became at least as important as verbal signals in the close quarters in which we worked—especially since talking was reduced for the sake of the wildlife, as too much talking can be stressful and cause imprinting: "The shift was filled with mostly new (human—I recognized some of the birds from my previous shift) faces, as well as new personalities. I was a bit surprised with the level of talkativeness of the workers in this shift, compared with my first shift. At orientation, we had been encouraged to not talk to the birds, and I assumed that this same principle would be in effect when talking around them, as well. This recalls another comment made by the orientation leader, to the effect that we should use technological devices as little as possible while at the WRC. She added that a past patient, a blue jay, had learned to mimic the sound of a telephone, and was subsequently euthanized."

Escape was a negative category that nonetheless frequently had to be dealt with—most often in the avian nursery. Nearly every avian volunteer session included at least one escape and hopeful recapture attempt. Such attempts were likely bound to be limited by the closed environment of the nurseries: "Some birds at this stage are already so good at escaping, however, that they will slip through what seem impossibly narrow openings, which usually necessitates catching them with a net, or bare hands if they are less agile. At least a half-dozen birds escaped from baskets I was tending today, which was frustrating, a bit embarrassing, and potentially injurious to the already-stressed birds. Although notes are often posted on the clipboards that accompany each basket, with helpful messages such as 'escape artist!' written on them, it remains difficult to predict an individual bird's behavior until I have tried to feed it at least once. Even after this initial exposure to a bird's tendencies, however, some are just too good (or bad, as the case may be)."

Adjustment was a ceaseless necessity and constant challenge. The unpredictability of the situations encountered provided many chances to test my ability to adapt to novel settings. As macro-scale adjustment was necessary for becoming initially oriented to each nursery, finer-grained, micro-scale adjustment was constantly being performed as well: "This is a job in which it is necessary to constantly 'shift gears' (or at least be ready to do so at a moment's notice)—not something I am particularly adept at. One is forced to adapt continuously to different species and their differing nutritional requirements, differing numbers of birds in each basket, the fact that some will take food willingly (even perching on the syringe, in the case of some very aggressive house sparrows [*Passer domesticus*]) while others will require the difficult and frustrating effort of force feeding, the ever-present threat of escape, etc. On the last point: I am heartened by the greatly reduced number of occurrences of escape that I directly contributed to last session (probably about half compared with last time). What this session lacked in number of escapees, however, was made up for by skill in remaining escaped."

Perfectionism was a stance I often found myself adopting. The completion of a perfect rehab session was an ideal I always held out. My tendency to obsess caused me to frequently second- (or third-) guess my actions. This was probably beneficial, overall, to the patients, but necessitated a great amount of energy expenditure on my part: "This week I was solely on cages for the first time. Unfortunately for me, cages are a perfectionist's nightmare. The birds in the cages are at a bit of a transitional stage, somewhere between the confined dependence of baskets and moderate independence of the flight room. Before birds can be sent to the flight room, they must be demonstrably able to feed without direct assistance (aside, of course, from providing them with dishes of water and food)."

A certain level of confidence was necessary, but at times hard to acquire. On the other hand, it was important to guard against hubris. Striking a balance between realistic confidence and hubristic overconfidence was both a challenge and a practical necessity: "Although I still see dehydrated (marked by visible lines on the abdomen and papery skin) and obviously ill birds, my perception is that they are, overall, healthier than when I began mid-June. This might be partially due to what I at least hope is an increased proficiency on my part in interpreting status and administering nutrition and hydration; or, given its unscientific, impressionistic basis, it could be nothing more than wishful thinking. In any case, I am beginning to feel more comfortable and confident in my ability, although I still find myself constantly glancing back at the feeding guide charts when I'm working on incubators (which I was this week)."

Death became an event of major significance. The lack of ceremony associated with wildlife deaths compared with human deaths was striking for me. The easy disposability of the animals prompted the question: Will a point be reached when the wild dead are treated with the same respect and dignity as human dead? "I was saddened to hear, at the beginning of this week's volunteering session, that a nighthawk had recently died. Nobody was sure what the cause was. As I shared in my last entry, I fed a nighthawk last week. I immediately began to question whether I had contributed to its demise. Given my complete lack of familiarity with this species, perhaps I had done something very wrong in handling and/or feeding it. I was somewhat relieved to find, upon examining the cages, that a cage with a 'common nighthawk' tag on it was empty (presumably meaning that this is the one that died). I quickly checked to confirm that the one I had worked with was still in the incubator, alive (it was). So, given that I do not recall working with any nighthawks in any cages, I think I can safely assume that I didn't directly contribute to this bird's death. Fine. But why do I care so much that I am able to prove that I had nothing to do with it? Does it really matter, for the bird, whether I, someone else, or some completely impersonal force caused its untimely death?"

A willingness to attempt to transcend my comfort zone was often necessary. Being willing, however, never turned into enjoyment for many tasks. Examples included scraping feces off linens, injecting squirrels with saline solution, and feeding and weighing angry squirrels that were ready to go to outdoor cages: "There was a plaque on the wall stating that the wearing of gloves and a mask is mandatory. I soon found there to be good reason for this, as many of the laundry items were covered in feces (among other vile things). The stench was astonishingly bad; I thought at one point that I might vomit. Some of the laundry had apparently been there for a while, as there were tiny maggots wriggling around in the feces. I could understand, now, why there were baskets of dirty laundry not only in the laundry room, but lining half of the adjoining hallway."

Contact with individual animals was a concrete reality that differentiated this project from many conservation efforts. When I entered a nursery, I received a distinct sense of a room filled with many individual organisms, but not a population. As time passed, I began to acquire knowledge of the behaviors and habits of many individuals. Never, however, was the realness of individual lives more apparent than with organisms I had rescued, as in the following case of the house sparrow I found in our alley: "Although we both wanted to give it water and food, it is generally recommended that baby birds not be offered food or water until they can be examined by a vet (unless, of course, for some reason they must be with you for a long time before this is possible). Since I figured that the probability of serious injury was high, we forced ourselves to abide by these guidelines and hope for the best in the morning. The bird appeared to be shivering, so we placed it (in its plastic container) in an empty mouse cage (so that it wouldn't be bothered by our cat [*Felis catus*], Squiggy, or dog [*Canis lupus familiaris*], Tonky), on top of the heating pad. After waiting a while and checking on it, it was still shivering, so I made it a nest and "blanket" with toilet paper, and loosely placed a lid over the top of its container so that it would retain more heat (while still allowing air in to breathe). "Petey," as we had named the bird, would have to tough it out for the night."

Vulnerability was a theme common to humans and nonhumans alike in this setting. While I was rendered vulnerable to attacks from patients, the patients were likewise vulnerable to my own often unskilled hands. A factor distinguishing my vulnerability from their vulnerability, however, was the conscious choice that was made to be vulnerable in the former case: "One rather disturbing incident I feel I should relate, but that I didn't think to include in the previous post, occurred in New Haven. Miriam, her brother Carl, and I were walking to a coffee shop. As we were passing Wooster Square, I noticed a lifeless-looking bird on the sidewalk. Before I was able to say anything about it, Carl stepped directly on it … I think this recent experience, though, illustrates the ease with which humans, both metaphorically and literally, can tread upon nature without even being aware of it."

Dying can be contrasted with death on many levels, including lack of finality, uncertainty, and suffering. Judgment calls were frequently made as to whether an animal was about to die, and whether this outcome was certain or avoidable. The process of dying was often hard to discern for even the most experienced staff. Surprise often accompanied the knowledge, after death, that an animal had indeed been dying before our eyes: "All seemed to be going well, save for a goldfinch who refused to gape and had to be force fed each round. At some point mid-session, I noticed a chimney swift (*Chaetura pelagica*) that an intern had presumably placed in one of the incubators. On its container was a note reading, 'Do not feed—drying off.' Assuming that it had simply taken a bath in its water dish and gotten cold, I left it alone for an hour or so. When I finally checked on it, it was obviously in bad shape—propped up against the side of the container and very lethargic. I noticed that one of its wings appeared longer than the other. I showed it to the interns, who agreed that it didn't look good. One of them said something about a vet check (which I don't think occurred). After asking whether I should give it anything, and receiving a negative reply, I put it in the warmer of the two incubators after confirming that this was okay. Shortly after this, an intern came over to inquire about its whereabouts, and to my surprise, it was no longer in the incubator! It turns out that another volunteer had, unbeknownst to me, removed it and was either force feeding it or giving it LRS (a rehydrating solution)—neither of which seemed a bad idea to me. The intern told the volunteer to put it back in the incubator, which she did. The next time I checked on it, it was moving a bit, and both an intern and I decided it looked a little better. The next time I checked on it after that, it was dead—stiff, eyes open, feet clenched."

Surprise became the rule, rather than the exception, of each session. As surprise became more anticipated, it also began to feel less shocking.

"Expecting the unexpected" helped lessen the potential extremeness of the experience of surprise: "Surprisingly, there were roughly twice as many patients in the incubators this week as last, the majority of which were American goldfinches (*Carduelis tristis*). Goldfinches, I was told, tend to nest late in the year, and my recent experiences at the WRC support this. Between the incubators and baskets, there were literally dozens of them this week."

Impressions were constantly evolving phenomena that usually took some reflective work to discern. They were highly challenging for me to sense as I was being bombarded with new information while initially learning about each nursery. A thorough analysis of the impressionistic content of my experiences was often only possible after the fact. After reflection, however, new access to impressions often generated valuable insights: "A general, initial impression of the mammal nursery is that, because of the far fewer feedings necessary per patient per session, more time can be devoted to assessing the condition of, and caring for, each individual organism if there is a full volunteer load. I was a bit surprised today to find four to five volunteers (including myself) at such an early shift (8 AM–noon), and on only the day after the opening of the nursery for the season. To see this kind of dedication is encouraging, and exactly what is needed if wild lives are to be spared in the increasingly hostile battle between wild animals on the one hand, and humans and their technologies on the other. It must be admitted, however, that such an analogy of war is only partially appropriate."

Procedures are the concrete outcomes of policies, and do not flow uniformly from them. This relationship revealed itself very clearly in different interpretations of rules or guidelines and resulting actions. How to most effectively proceed from "A" to "B" often boiled down to judgments of individual crew leaders, which were sometimes conflicting: "Recordkeeping procedures were markedly different today, and possibly more organized. Rather than each volunteer taking a book in which to record weights of squirrels, amount fed, strength of formula, presence or absence of diarrhea, etc., individual files were removed from the three-ring binders pre-shift by coordinators. When all the duties pertaining to a particular basket or cage are completed and every relevant field filled in, the files are placed back in the binder (hopefully in order—the first time I attempted this, I put the file on the wrong side of the marker tab)."

Violence was a deceptively common reality in such a care-focused setting. Not all violence was harmful; some, such as injection, was needed to effect healing. This violence with intent to heal, engaged in by human volunteers, can be contrasted with the more conventional understanding of violence as being driven by intent to harm, which was apparently engaged in by some patients. Many actions that were initiated with a sense of calm passivity, such as feeding, escalated into violent attempts to control and/or escape: "I fed and cleaned cages for several particularly aggressive self-feeders for the first time last Tuesday. Upon approaching the cage, I was greeted with growling and was happy to be wearing thick gloves as I was bitten numerous times while trying to get hold of them. As many of these cages have four squirrels in them, often with different colors of nail polish on their ears that must have been mistakenly applied, it is extremely difficult to both identify and capture them for weighing, as well as keep any from escaping. Unfortunately, I had numerous escapees last Tuesday, with one particularly aggressive squirrel seeming to jump at my face from the head-height cage and, despite my attempts to catch it, fall to the floor."

Release is a desired end goal of all rehab efforts but is not always actualized. Moreover, guidelines for being released, as well as release practices, varied greatly according to context of admission and type of animal. Conflict was often displayed over proper duration of stay prior to releasing. Although release may be perceived, overall, as a time of ease and freedom, it is often not without its own stress, as the following account of a squirrel release shows: "I was a bit nervous about opening the cage, since I had no gloves, and expected a mad dash to exit the cramped quarters. Surprisingly, however, the grays did not leave immediately, or even seem to want to leave. We tried taking the wooden container out of the carrier, tipping it forward, holding it upside down, even shaking it, but the only result was the emergence of nine—yes, nine—bushy tails. I thought about trying to take the box apart but decided it would be better to just wait it out, hoping that the braver and more inquisitive of the inhabitants would make their way out given some time. This is just what happened, but slowly."

Success was harder to define than I expected. I came to find that a successful session was often marked more by a personal perception than by "objective" data, such as might be provided by, for example, post-release monitoring. Any knowledge or feeling of success was usually only temporary and partial: "According to a whiteboard in the hall outside the squirrel room, the WRC has now released its 200th squirrel this season! This is exciting, especially since the season is only half finished. A question I have found myself struggling with in the past is raised by this number, however: Is it an indicator of a successful beginning to the season, or of tragedy? Something to be celebrated, or regarded with resignation (or even mourning)? After all, the success rate for rehab centers hovers around 50% (although I don't have specific data on the WRC). This percentage, moreover, does not consider post-release survival. So, were the 200 released the fortunate members of roughly 400 admits, half of whom did not survive? I do not know, but I am interested in finding the answer. What is more, how many of these 200 will survive more than a month in the wild? 100?"

A sense of newness was nearly always present, even in such an ordered (at least in relation to nature) setting. This is partly because of the obvious fact that new patients were constantly coming in. However, it was also due to the mixed novelty generated by the necessity of performing multiple tasks in multiple nurseries and other areas. Although some aspects of the nurseries—especially the mammal nursery—resembled an assembly line, the "assemblages" being treated were far from inert and processed products: "Two new 'things' this week: shift leader, and Virginia opossums (*Didelphis virginiana*) (in addition to the semi-new pox birds). It is beginning to seem as though expecting the unexpected is warranted for the so-called squirrel nursery. The new crew leader said that she is technically a co-crew leader with the expected person but had been unable to work in the nursery until now because of summer classes (she is a pre-veterinary student in the University of Minnesota's Animal Science program)."

Rescue is the necessary—and often hidden—first step toward all rehab efforts. A relevant distinction for rehabbing is that of perceived versus actual rescues. Young animals, in particular, are often brought to the WRC with the belief that they are orphans being rescued, while they are in fact healthy individuals being removed from their parents and habitat: "As Tonky goes over to the well often to sniff and look down into it, I did not think much of her behavior at first—except that this time, she would not leave. I decided to look myself and was surprised to see a live baby rabbit in the well, about the size of my hand. I at first thought I might be able to scoop it out with a shovel, but then realized the well was too deep and the rabbit too mobile. It appeared that I would have to lower myself down into it, which is what I did, being careful not to scrape myself too hard on the broken plastic edges. I just fit."

Observing was more clearly and easily distinguished from participating early on. For example, a short time of nearly pure observation marked initial volunteer efforts in each nursery. As participation became more routine, it became easier to both perform tasks and neglect to critically observe. The following is taken from preliminary observations of the waterfowl nursery: "The layout of the waterfowl nursery is very different from that of either the avian or the mammal nurseries. The first and most noteworthy difference is that water is everywhere, compelling the wearing of rubber boots. A hose with a high-pressure spray nozzle is perhaps the main instrument of the room and is nearly always in use by someone for spraying the floor, pre-cleaning dishes, or filling water troughs. The young birds (not all are ducklings—indeed, not all are even waterfowl) are kept elevated above the floor in long, open-topped wooden boxes covered by mesh fabric, and are equipped with heat lamps and suspended feather dusters, which the birds flock to if they feel threatened. Each one of these rectangular boxes is occupied by from one to about a dozen birds."

Wings became a frequent reminder of the fragility of both life itself and attempts at healing. The status of being winged, usually flighted creatures helps account for lower recovery and release rates of birds compared with mammals. The "angel wing" epidemic among waterfowl was probably the most serious systemic, unsolved problem I saw at the WRC: "I visited the outdoor waterfowl cages for the first time today. The mallard cage is a large enclosure directly outside the nursery on the south side of the building, with plastic swimming pools instead of smaller tubs, filled with water and sprinkled with duckweed. I filled three large, circular plastic feeding dishes about halfway with feed pellets, and placed one in each section of the outdoor mallard cage. These are the 'starter' pellets; perhaps they provide adequate nutrition for very young birds, but not juveniles (or at least not as most of their diet). Could this be part of the explanation for the mysterious duck angel wing epidemic?"

Care was transformed from an empathic feeling to a physically performed set of actions. A significant change occurred in the transition from caring *about* something to caring *for* something. As the following incident indicates, however, active caring is often rife with ambiguity and entails risk that can have negative consequences: "An incident of confusion, ending in unfortunate news, occurred Sunday. A woman in scrubs, who I believe to be a vet tech, entered the waterfowl nursery asking if I knew of any woodpeckers there. I said that I did not, and we both thought it strange that any would be in the waterfowl nursery given they are normally kept either in the avian nursery or in an outdoor cage. She left but came back later after realizing that she had misread some messy handwriting that referred to a wood duck. The crew leader then entered the nursery and, after being questioned as to its whereabouts, replied that she had found it earlier on its back and hypothermic. She had placed it in the oxygen room, thinking this might help it, but unfortunately it was found dead shortly thereafter."

Solutions, while often hard to come by or to know with certainty, were in some ways easier to identify and enact than for more abstract pursuits. This is to say that the solving of problems could be empirically affirmed in a way that is often not possible with conventional conservation—through direct observation of individuals. However, complete certainty as to recovery was never achieved due to lack of post-release monitoring: "Deciding it was time to do something about this death trap (egress window well), I rummaged around in the garage for something to put over the large opening but found nothing suitable. I called my landlord, and he was there the next day, putting new covers, alas, on all but the problem well (that he at least was covering the smaller wells was good, however, since I had seen many rodent skulls in them). I explained that the large one was the most problematic, and needed covering the most, but he replied that he could not put any permanent or heavy coverings on it, since the basement resident needed a means of escaping directly from his unit in case of fire. The best resolution we could come to was that we would think about a solution. Until one appears, I will try to monitor the well closely."

Role was constantly being worked out in the nurseries. While roles tended to overlap, some volunteers were more comfortable or willing to perform some tasks than others. For example, even though I feared feeding "self-feeding" squirrels, I was often asked to perform this task for cages with very aggressive patients: "As I was cleaning the first of the indoor divided enclosures, the girl, seemingly taking on the role of the intern in her absence, told me that I should not bother removing the food and water containers as I had just done, since they could just be refilled while in the enclosures. I had assumed that both would need to be washed, but then remembered that the large, heavy water dispensers are often left where they are and filled with a hose (the food dishes, on the other hand, certainly needed to be washed, as ducks often use them for toilets as well). I was a bit annoyed at being bossed by a teenager who seemed to have little respect for my seniority of age and degree, but remembered the special context I was in, where textbook knowledge must often bow to the dictates of direct experience and established procedure."

Technology was perhaps the most apparent factor distinguishing this setting from most other settings in which wildlife is found. Of course, technologies pervade places urban wildlife reside, as well. But in no other situation are wild animals so completely insulated from nature solely for their own good: "We have apparently been blessed with a bounty of duckweed, as I have not seen any lettuce substitute recently and the cat litter turned duckweed container seems full most of the time. Unfortunately, it was also frozen today, and I scraped skin off my knuckles trying to chip away at the frozen mass with a screwdriver. It would be interesting to know what proportion of objects at the WRC have been repurposed from intended human uses; my guess would be the majority. Although I am aware of a substantial literature on the unintended consequences of technology, I am unsure whether much has been written on unintended uses of technologies. The WRC might be a good place to start such a research program."

Technique was a procedurally critical piece to every interaction. Both "technique" and "technology" derive from the Greek *techne*, roughly meaning "craft." While some tasks, such as giving meds, were more inherently technical than others, the WRC can be broadly viewed as an environment permeated by *techne*: "I held a wood duck (aka "woody") duckling for the first time today as it was being re-banded. The strength and vigor of struggle this little bird displayed was unexpected, and I nearly dropped it. One of the interns told me that the ducks must be re-banded frequently for purposes of identification, as they grow quickly. If the band on the leg is too tight, it can cause circulation difficulty as well as discomfort. A rule of thumb for band replacement is if the band cannot rotate freely on the leg, a larger one is needed."

While relearning was necessary after the intervening time between volunteering seasons, it was also needed to a lesser extent during every session. Memory work was an ever-present necessity. Learning was never a final place to be arrived at, but understandings of situations and problems were constantly being refined, redefined, and renegotiated: "I am afraid that I have lost my touch a bit with the syringe, and made several squirrels aspirate as a result of this combined with their eagerness to feed."

Guilt was an emotion I felt after failing to perform as I had desired. A guilty conscience from arriving late, missing a shift, or taking poor care of an animal was sometimes enough to prevent repetition of the error. In more than one instance I experienced guilt after learning of the death of an animal I thought I had treated. I also found myself wondering whether the veterinarians felt guilty after performing euthanasia: "I did not volunteer at the squirrel nursery on Tuesday, since I was not on the regular shift list for this date. I also was delinquent on Thursday; I justified this by my feelings of illness that morning, as well as my memory of the crowdedness of last Thursday's shift with eight human bodies in the nursery. Recalling that Thursday's shift leader had requested help on the Friday morning shift, I made it to this one. Upon entering the squirrel nursery Friday morning, however, I was very sorry that I had not showed up Tuesday and Thursday."

Infection is a term I was frequently confronted with but did not fear much. Squirrels were constantly battling respiratory tract infections, and those of us (such as me) who had not been vaccinated for rabies were reminded often to not handle animals other than birds and squirrels. Not everything deemed infectious was negative, however, attested to by the infectious enthusiasm of many volunteers and crew leaders: "A spate of respiratory tract infections (RTIs) seem to have a death grip on the squirrel nursery right now, and I have noticed a significantly greater number of DIC (died in cage) and E (euthanized) designations than I recall for the first litter of the season. Roughly every other squirrel I pick up appears to show the symptoms of an RTI, by way of wheezing, sneezing, and/or mucus around the nostrils. Many are on medication or probiotics."

Memory was crucial for both the successful and efficient completion of tasks, and accurate journaling after each session. Conscious attempts at memorization were not often engaged in, but recall was aided by routine and consultation of crew leaders and reference materials. Perhaps the most consistently challenging thing to remember that required a great amount of accuracy was the very short term, yet highly repetitive association of particular individuals (as indicated by nail polish color) with their respective weights: "Given the relative lack of outstanding experiences this week, coupled with the nearing of the end of my documentation period, this seems like an appropriate time to reflect on some of the things I have learned thus far. Thinking back on the beginnings of my interest in wildlife rehab, where has this path of inquiry and discovery taken me? This is a broad and difficult question to answer thoroughly. So much of what I have done has been lost to memory already, or never even brought to the level of consciousness to begin with."

Causes were often difficult to pinpoint, especially for systemic problems such as angel wing. Although delineating causality is not the primary goal of rehab, it is nonetheless helpful for acquiring knowledge relevant to both healing and learning. A special challenge of rehabbing is that human patients can often verbally tell what has caused their problems, while wild animals cannot: "A question that has been on my mind, and that has come up in discussion during shifts, is the cause(s) of and health status of the second yearly litter, in general. Is its production merely an evolutionary tactic to buffer the potential catastrophe of a failed first attempt at breeding for the year? If this is the case, why is the 'strategy' not more common among other species and taxa? Or perhaps it has nothing to do with selection or evolution and is simply a random occurrence."

Promotion was a surprising event for me. The prospect of being promoted to crew leader never seemed real for me until it happened. This change fundamentally altered both my role in the mammal nursery and my relationship to other volunteers: "Today was a trying day, the first day as a (co-)crew leader. I was considerably more anxious than normal before the shift, and my nervousness was only compounded by the necessity of telling my fellow crew members that I was now one of their crew leaders. Did they take this as an indication that their work was somehow deficient compared with mine? Perhaps they felt as though they deserved the position more than I did, lacking in tangible benefits though it may be? I am not sure that drawing crew leaders from the volunteer pool late in the season reflects good policy, because of the possibility of hurt feelings and resentment."

Illness is often fraught with ambiguity, from an external viewpoint, in a way that the other categories of reasons for rehab are not. It is, however, rarely ambiguous for the one who is ill. At one point during the second season, I became violently ill, and nearly went to urgent care. I was never sure if this illness was caused by my animal patients. While difficult to endure, being ill helped me to relate to the plight of many of my patients: "I became rather violently ill last weekend with flu-like symptoms, and so unfortunately, I missed my Tuesday shift. I wonder whether this illness is linked to the increased anxiety and stress levels experienced performing my new duties as crew leader. Alternatively, but relatedly, perhaps I contracted something from the squirrels or humans in the nursery. Squirrels are not known as being particularly zoonotic creatures; but I do recall the volunteer coordinator at the WRC having recently been out sick for what seemed like a relatively long period of time, and another volunteer stating that she thought she had West Nile virus."

Disorientation was a feeling initially present at first that subsided (but never completely vanished) after a time. The formal orientation session that preceded volunteering certainly helped guide me. However, the number of variables present at the WRC was mind-boggling, at times: "From my recollection, the difficulty began when I stuck myself for the first time with a needle. Luckily, I did not bleed this time. However, not long after this first incident I stuck myself again, and this time I bled profusely. Remembering where the first aid kit is kept in the treatment room, I disinfected the wound with an alcohol pad and placed an adhesive bandage on my finger (I learned afterward that I should have logged this in an incident report, which I do not recall having previously heard about).

After this, while drawing up meds, I absentmindedly drew several syringes far too full, having misinterpreted the hundredths as the tenths place. Luckily, my co-crew leader caught the mistake, and questioned by asking, "Who are those for?" Not remembering the case numbers, I replied, "Squirrels," after which a crew member quipped, "You'd make a great detective!" This incident put me in a spiral of second-guessing that lasted the rest of the shift. I felt powerless, knowledge-less, unable to make decisions. I felt as though every comment and question directed at me now was an attempt at domination, all the while trying to remind myself to not take things personally and wondering how accurate my perception was of these messages as motivated by distrust or a desire to discredit."

Identification with those in need, though "located" across the species line in my case, increased with time spent in the nurseries. This ability to identify is closely linked to the experiences of sympathy and empathy. I did not write much about sympathy, empathy, or identification in the journal, but found myself thinking about these themes often. The position I took was more often closely aligned with sympathy than empathy, but, as was mentioned, becoming ill helped me to more accurately imagine the plight of my patients, and to identify more with common characteristics such as suffering and susceptibility to illness.

References

Bryant, A., and K. Charmaz. 2007. Grounded Theory in Historical Perspective: An Epistemological Account. In *The SAGE Handbook of Grounded Theory*, ed. A. Bryant and K. Charmaz. London: SAGE Publications Ltd.

Daiute, C., and C. Lightfoot. 2004. *Narrative Analysis*. Thousand Oaks: SAGE Publications, Inc.

Hindmarch, I. 1972. Observer Participation. *Drugs and Society* 1:6–10.
Jorgensen, D. L. 1989. *Participant Observation: A Methodology for Human Studies*. Thousand Oaks: Sage.
Kidd, A. H., R. M. Kidd, and R. L. Zasloff. 1996. Characteristics and Motives of Volunteers in Wildlife Rehabilitation. *Psychological Reports* 79:227–234.
Peggs, K. 2012. *Animals and Sociology*. Hampshire: Palgrave Macmillan.
Pesic, P. 1999. Wrestling with Proteus: Francis Bacon and the 'Torture' of Nature. *Isis* 90 (1): 81–94.
Ross, L., and R. E. Nisbett. 1991. *The Person and the Situation: Perspectives of Social Psychology*. New York: McGraw-Hill.
Stocking, G. W. 1960. Franz Boas and the Founding of the American Anthropological Association. *American Anthropologist* 62:1–17.

CHAPTER 12

Discussion and Conclusion: From Metaphysician to Physician

Abstract While any conclusions to be drawn from such an exploratory project are, indeed, bound to be partial, my hope is that an attempt at summarizing some of the most important aspects of this book will help to seed future discussions about conservation issues. In very broad and general terms, this has been a project of truth-seeking. A more contemporary approach to framing such a project might instead focus on the attempt made to begin building a bridge of agreement between two related, but sometimes conflicting, fields. Regardless of whether the foregoing is interpreted as more objectivist or subjectivist, however, it should be understood as constituting a step toward placing the theory of conservation biology on a surer philosophical footing.

Keywords Animal welfare • Metaphysics • Biology • Analogy • Population

While any conclusions to be drawn from such an exploratory project are, indeed, bound to be partial, my hope is that an attempt at summarizing some of the most important aspects of this book will help to seed future discussions about conservation issues. In very broad and general terms, this has been a project of truth-seeking. A more contemporary approach

L. B. DeVaney, *Animal Conservation Ethics and the Population Problem*, The Palgrave Macmillan Animal Ethics Series,
https://doi.org/10.1007/978-3-031-91891-9_12

to framing such a project might instead focus on the attempt made to begin building a bridge of agreement between two related, but sometimes conflicting, fields. Regardless of whether the foregoing is interpreted as more objectivist or subjectivist, however, it should be understood as constituting a step toward placing the theory of conservation biology on a surer philosophical footing.

Philosophically, I have attempted to clear a path toward a view of conservation that is more sympathetic to individual animal welfare. This has been accomplished both rationally and empirically. The vision I aimed to achieve, moreover, was strongly influenced by conscience and moral sentiment. Hopefully, a conspicuous gap in theory has been filled adequately.

Why bother with such a project for the applied, practice-oriented field of conservation? To the extent that such theorizing is or was in fact necessary, has it not already been done by the likes of Emerson, Muir, and Leopold? My answer to such fitting questions is that the nature of the discipline, along with the wider world itself, is changing, and that this change is beginning to ask for a new view. This view, however, represents an ideal that itself does not change—or at least not at the same rate as the real world. The goal that this ideal points to is that each individual organism should be able to live a life of flourishing.

From a certain perspective, our modern era, often portrayed as a time of insurmountable environmental crisis, can be seen in a more positive light than it is currently being cast in. I have attempted to provide the beginnings of a view from which an exchange of values might be made, to enable the growth of this more positive perspective. I will now briefly retrace the steps taken to get here.

This book began with an outlining of the most important points of my stance. From here, the main body of the work was launched into. The body was divided into three parts, reflecting the distinctions and connections between theory and practice, and an ideal chronological sequence of thinking and acting. The first part centered on the meaning and metaphysics of conservation biology as an academic field. This was seen as a proper starting point because of the lack of attention that has been given to investigating the assumptions inherent in conservation biology, as evidenced through both the terms commonly used in the field and the types of actions often engaged in. Rather than beginning with ethics, a popular point of departure in environmental philosophy, this project sought to begin from the rock bottom of metaphysical thinking. A central focus of the first part of this book is the meaning of conservation and biology as understood both apart from one another, and together as a presumed unified discipline.

The meaning and proper understanding of "whole" is of special concern here—and is, indeed, a theme that runs throughout the work. In particular, the concept of super-organismal natural entities (SONEs) is introduced to approach and discuss the issues of scale and size as they pertain to conservation. While the large and complex wholes represented by SONEs are usually presumed to be the entities of proper conservation concern, questions are raised as to the appropriateness of this current focus on what are often deemed systems, as well as to their metaphysical status. Central to this analysis are the basic classes of things and stuffs. A treatment of ontological categories is entered into to help distinguish between aspects of existents that are relevant to their classification. Bounds and unity are identified as concepts of relevance for biotic classification which are often found to be lacking in SONEs.

Questions about the extent of conservation biology's identity as and relationship to biology proper are asked. Among these are whether conservation biology is to be understood as a special type of or area within biology; conservation applied by the field of biology; or conservation applied to the field of biology. Problems are found with each of these conceptualizations, leaving the question of how, exactly, conservation biology is biological largely open. A possible way out is identified: By eliminating the "biology" term, and perhaps replacing it with "biotic" (as in biotic conservation), confusion may be reduced. This might do away with the implicit indication that the study of life is the main goal of conservation, exchanging a discipline-neutral word for a discipline-specific one, and more clearly identifying conservation itself as the desired end.

The possibility, however, that conservation biology may be viewed as a special type of biology is not entirely done away with. Aristotle's four causes are enlisted, in hopes that an advancement of understanding might be reached through an increase in conceptual resources that might save the traditional framing. An investigation is made of the appropriateness of conceiving of conservation biology as final biology. Final biology, while potentially meaning many different things, can be broadly understood as a stage or phase of biology where the study of life is increasingly accompanied by, and eventually replaced with, practices aimed at life's preservation and care. Put as a question, might a saturation point in "normal" biology be in the process of being reached? This could be framed as a paradigm shift. If such a conception is valid, however, it brings up other issues regarding the continued integrity of biology as a unified discipline or set of disciplines.

Particular attention is paid in this work to the level of the population, which is acutely focused on by both conservation and other biologists and rarely questioned. Are populations real, mind independent features of the natural world, or do they merely represent a convenient way of thinking? A move is made from the smaller and simpler physical categories of energy and matter, which are held to pervade the entire universe. Progress through increasingly larger and more complex aggregations and entities is achieved, until the levels of organism and population are reached. It is found difficult, however, to begin to understand vertebrate populations as either concrete stuffs or things. With the priority often given to populations in conservation laws, plans, and actions, this seems troubling.

While the definite, absolute "thing-in-itself" of the population remains elusive, it is hoped that a better understanding may be achieved through apprehension of purported attributes of populations. Although some progress is achieved with different heuristics, these heuristics are found to be largely metaphorical and, therefore, not composed of literal content. The further, broader question is raised of the extent to which it is possible and likely for the use of metaphor and analogy, and mental images in general, to provide an enhanced, accurate understanding of natural, external entities and phenomena.

Metaphor-mediated understanding of ecological concepts is, by and large, found lacking in clarity and rigor. For example, problems are uncovered with the commonly used ecologically oriented notions of growth, stability, equilibrium, and the purportedly unifying ecological picture of the "layer cake" (Allen and Hoekstra 1992). In being aimed at a new target, have such metaphors drifted so far from their source domain that they have lost their meaning? The potential for having committed a category mistake is discussed in connection with these and other metaphorical concepts.

Although the foregoing critical analysis uncovers difficulties with many received ideas in conservation biology and ecology, a potential way out is offered through the positing of a softer interpretation of the meaning of SONEs. For example, it may be the case that SONEs should be understood not as physical things or particularly cohesive systems, but as a politically helpful way of demarcating aggregates with statistics and other methods. This "soft" view, juxtaposed with the criticized "hard" view, shows promise as an approach for helping avoid the commitment of a category error when thinking about populations. An effect of taking such a view, however, is that justifications for population-level (and perhaps larger scale) conservation actions are significantly weakened.

Why might such an approach be better than simply ceasing to refer to populations at all? The short answer is that much rhetorical force can be found in reference to populations, species, ecosystems, and the like. This language has been historically used quite successfully—especially in the United States—to foster protection and preservation of large areas all at one time, thereby obviating, in the minds of many, the need for initiatives directed toward smaller entities such as individual organisms. Conflicts begin to arise when benefits granted to species, subspecies, or populations cause costs or risks for individuals. Some SONE categories are more realistic than others, however, in their account of entities and their relationships. For instance, the community concept does a better job than either the population or species concepts of highlighting both the relative freedom and autonomy of adult vertebrate animals, and the important and meaningful social bonds that arise in the contexts of both intraspecific and interspecific communities. While a community is clearly not a physically unified whole to the extent that an organism is, the claim that a level of social accord exists within many communities looks to have merit.

To give a more specific and detailed picture of what can happen when populations are framed as things—but in this case treated otherwise—the discussion progressed to the contemporary case of gray wolves in Minnesota. This can be seen as a shift from a metaphysical to an ethical treatment of conservation issues, where argument begins to bridge the gap to practice. Although the putative unit treated in recent legislation is referred to as the Western Great Lakes Distinct Population Segment (DPS) of wolves, both observation and the practice of killing easily show that this group of wolves does not constitute a truly distinct, discrete, unified segment. It seems that wolves are being portrayed as one thing by the law and "science," and another by the senses.

A brief natural and legal history of the gray wolf, at both state and wider levels in the United States, is given. Most notably of interest is the 2011 federal ruling to remove Endangered Species Act protection from the Western Great Lakes DPS,[1] coupled with the instatement of Minnesota's first state-sponsored wolf hunt in 2012. Through this process, government prompting to think in terms of one category but act in terms of another—in this example, to assent to unity while both seeing and encouraging disunity—appears to have been conveyed to the citizenry. In addition, it is deemed that insufficient weight was given to public opinion in the preseason survey administered by the Minnesota Department of Natural Resources, and that the season itself was poorly monitored and managed since the kill quota was exceeded. While it was previously acknowledged that a soft view of populations may have certain positive implications if purported population-level concerns are not taken to trump organism-level concerns (i.e., if populations are seen and treated as aggregates of individual organisms), it is argued that neither this view nor treatment were taken in the above case. On the contrary, in this case massive confusion has arisen and been propagated by state and federal governments for the apparent sake of short-term profit gains.

The book proceeds by building on the metaphysical foundation of the first part and continues the transition into an ethical perspective. This transition is necessarily continuous, with boundaries between the philosophical branches rarely discrete or clearly made. Given the apparent incoherence resulting from attempts at applying basic moral principles such as "do no harm" to objects of common conservation concern, are attempts at developing a conservation ethic a lost cause, despite conservation biology's claimed status as a value-driven discipline?

While the logical error of the naturalistic fallacy is often cited in reference to environmental ethics, the meaning of a more extreme version, referred to as the conservationist fallacy, is discussed. The naturalistic fallacy centers on the problem with attempting to derive an "ought" from an "is," whereas the conservationist fallacy extends this to the attempted derivation of an "ought" from a "was." It is pointed out that commitment of this error is frequent in ecological contexts, particularly regarding non-native species. This opens the discussion into treatment of other important conservation issues, such as scarcity versus abundance, and preservation versus restoration.

A question of implied relevance, how the differing emphases of ecology and individual animal welfare might be reconciled, is now explicitly raised.

[1] By December 2014 and after three consecutive hunting seasons, wolves had been relisted under the ESA (Erb and Humpal 2020).

A problem treated by J. Baird Callicott, "Callicott's triangle," is discussed as an example outlining some of the difficulties involved in not only trying to reconcile neighboring disciplines, but simply gaining an adequate understanding of the positions taken by each. More specifically, the views of ecological holism, humane moralism, and ethical humanism are described as instances of moral stances taken that show a nonlinear, triangular, and therefore likely incompatible, relationship to one another. I argue, however, that while differences exist among these views concerning levels of inclusivity and aggregation, conservation biology and wildlife rehab are more moderate and tractable approaches that have the potential to draw humane moralism and ecological holism toward a more unified response to ethical humanism.

A thought experiment is provided as something of a transitional piece between ethics and practice. Thought experiments can be helpful for abstracting from the many contingent particulars inherent in a situation and focusing attention on the most relevant phenomena—especially when such situations are shrouded in secrecy and therefore would be difficult to obtain data about. This is not to say, as the section on participant observation would seem to challenge, that particularities are not valuable or that they always demand abstraction. The "alien incident" is essentially a small set of scenarios, wherein an attempt is made to bring into clearer focus the implications of similar but significantly different circumstances.

The goal of the alien incident is to portray, in compact format and easily understandable terms, the role of core conservation concepts framed more realistically than is often the case, but from a fictional third-person perspective. Rather than simply asking "What would I do in such and such a situation?" which would be likely to hinder creative consideration of possibilities, the reader is prompted to consider, from a more objective perspective, what someone holding values commonly held in the ecological sciences would do. In the presented scenarios, however, it is worth noting that the participant's values do not change; what changes are the relevant parts of the situation and the participant's response to these situational items.

Included in this set of concepts are previously treated issues such as intervention, nativity, rarity, and sentience. The teaching strategy adopted may be an improvement over other heuristic approaches, including those mentioned in this book, in that the outcome is explicitly framed as fictitious. This "virtual observation" provides a sort of compromise between a highly abstracted, theoretical, and distanced perspective, on the one hand, and the total physical immersion experienced by the participant observer.

At this point in the book, a contrast is made between theory and practice. This is only to indicate that the first two parts of the work, on balance, are focused more on theory, while the third part is focused more on practice. A brief orienting overview of modern international and national wildlife rehab is provided, beginning with the hospital of Albert Schweitzer and progressing to the forming of the International Wildlife Rehabilitation Council and the National Wildlife Rehabilitators Association. Some basic guidelines for appropriate conditions of wildlife rescue are also given here.

Information of both global and local relevance is provided in the form of interview summaries. The results of in-depth interviews with the directors of two very different rehab centers are presented. My main objective in conducting interviews was to present an aspect of the practice of rehabbing that could not be adequately captured from a (non-director) volunteer perspective. While a picture of an archetypal general facility is constructed through Phil Jenni's representation of the Wildlife Rehab Center of Minnesota (WRC), the status of another Minnesota center, Wildwoods, as relayed by Peggy Naylor, serves as a juxtaposition of a growing home-based nonprofit seeking to maximize connections and available resources. Wildwoods can be seen as a microcosm of the challenges and opportunities faced by many private rehabbers, and the relationship of Wildwoods to WRC as an example of the potential for convergence, collaboration, and networking.

An attempt is made to enhance the visible connection between conservation and rehab by pointing out a way in which a currently recognized conservation problem is both shared with, and may be treatable through, rehab. This is the problem of data deficiency, and it is an epistemic issue. Conservation biology is primarily data deficient through lack of data on population or species numbers, while wildlife rehab is primarily so due to lack of data on post-release success.

The proposed solution to this problem may be less than complete, given its currently metaphysical, non-epistemological nature. Another way of putting this is to say that expanding rehab efforts may help us solve some conservation problems before necessarily showing that they have been solved. For instance, Arizona gray squirrels are currently listed by the International Union for Conservation of Nature as data deficient. While having a rehab facility near the squirrels' range in Arizona and New Mexico may help both individuals and putative populations, the precise extent of this aid may not be known for various reasons, including both lack of reliable number estimates and lack of post-release follow-up. The question is: To what extent does this (or ought this) privation of data matter? Can we—should we—be content to attempt to conserve without knowing or trying to know all the details? Might we rightly reframe data deficiency as data sufficiency, in many cases?

There is a sense in which wildlife rehab is and will likely remain inherently data "deficient," in that it prompts a rejection of the common scientific drive to transform and reduce the world to bits of information. This can be seen in its treatment of real, concrete, individual organisms as opposed to graphical representations and metaphorical understandings of SONEs. This is not to say that rehab is, will or should remain a practice completely divorced from data. On the contrary, the potential for collecting data on a host of environmental factors is great. However, the primary goal of rehab will likely remain the welfare of individuals, whether of scarce or abundant species.

The final chapter of this book begins with an account of the qualitative social science research method of participant observation. My main goal in portraying my time at the WRC through the lens of participant observation was to go further than making the simple claim that I had progressed from theory to practice. I wanted to show that there is a methodologically rigorous way of attending to a culture or subculture, whereby salient features of it can be uncovered. This was important for gaining a clearer picture of the small-scale conservation practice of wildlife rehab.

Participant observation is defined by seven core features, including development and adoption of an insider perspective; location in often mundane, everyday life settings; stress placed on interpretation and understanding; an open-ended, flexible, opportunistic logic of inquiry; in-depth, case study approach and design; the performance of a participant role or roles that involves establishing and maintaining relationships with natives in the field; and the use of direct observation as well as other methods of information-gathering. The time I spent as a volunteer at the WRC was found to have fulfilled each of these criteria but differed significantly from conventional participant observation in one crucial respect: An emphasis placed on a setting of both human and nonhuman existence. Because of the interspecies community found in rehab centers, this chapter falls under the rubric of anthrozoology, also known as human-animal studies.

Finally, I provide a summary of field notes and journal entries made during two seasons of participant observation at the WRC. As time passed,

I found my entries becoming longer, on average—contrary to my belief that I would have less to say after an initial learning period. I never ceased to be surprised by the novel situations I found myself in. This novelty was undoubtedly contributed to by my roles in three different nurseries: Avian, mammal, and waterfowl.

This book represents the beginning of a view from which the links between two life-oriented disciplines can be seen. These disciplines, while alike in many ways, currently ascribe to some notably different views on appropriate conduct toward wildlife. I have shown, however, that some of these differences are not entirely real but merely apparent, and that greater agreement soon is both plausible and desirable. To the extent that the practices and perspectives of rehab and conservation are in real disagreement, metaphysical problems can be found. Ethics, in my view, always boils down to metaphysics. I will now briefly offer some closing thoughts on how the association between conservation and rehab might be reframed.

As has been discussed, a current issue of major tension between conservation and rehab is scale. While many ecological scales have been posited and emphasized, species and populations have been of primary concern for conservation biologists since the official birth of the discipline in the late twentieth century. A major justification for this has been the position that large scale biotic entities or aggregates should be preserved for the long-term evolutionary benefit found in their genes. Wildlife rehab, however, has been focused on the next “rung” down on the ecological ladder, at the organismal level. According to many rehabbers, small scale entities, namely organisms, should be repaired or restored, if possible, for the welfare of the animals themselves and their interest in living. However, since populations and species are essentially groups of organisms, there is no apparent incompatibility between restoring individuals through rehab, and protecting entire populations. In addition, skillful rehab provides the benefit of easing suffering and improving animal welfare through treatment of individuals. As the organism comes to be seen as a true biotic whole, the scale of activities deemed environmentally conservative will gradually shrink.

Ecologists and conservationists seem to be nearly silent as to the issues of sentience and pain. This silence largely comes out of the admirable conviction that, regardless of whether members of a species are aware or can feel pain, if the group is endangered, then we should protect its members. Animal welfare does not seem to be on the radar of most ecologists, however, and since most conservationists focus on aggregations of individuals, welfare is not of preeminent concern for many of them. Since many Western ethical theories begin with the sentient individual, and people from diverse backgrounds strongly value animal welfare, this lack of attention to welfare from a self-described value driven discipline seems odd. Some leading figures in nature ethics, as I have noted, have argued against the compatibility of environmental ethics and animal ethics. Wildlife rehabbers, however, pay close attention to the pain and suffering of even common animals, and go to great lengths to give them care. Much recent research, moreover, has been directed toward erasing any lingering skepticism that animals feel pain. These claims of painience have been extended to fish (Sneddon 2020) and even some invertebrates (Crook 2021).

Accordingly, the compassionate conservation movement is beginning to increase the appeal of welfare as a valid conservation concern vis-à-vis an increase in empathy for wild animals (Heister 2022: 8; Lynn et al. 2025). One approach being taken via compassionate conservation is by arguing for the status of personhood in nonhuman animals. There is still much debate over the necessary and sufficient criteria for attainment of personhood. If animals are viewed as persons, this will go some way toward defeating the human exceptionalism that plagues our treatment of them (Wallach et al. 2020).

The related issues of awareness, sensation, pain, suffering, and consciousness open up a potential avenue for wildlife rehab to be considered, instead of an illegitimate attempt at conservation, a subtype of conservation where sympathy and compassion are both relevant and needed. It is hard to empathize with invertebrates and plants, but their abundant existence is necessary both as a resource and because of their status as living beings with the goal of perpetuating their kinds. So, while plants are very important, the special status of vertebrate animals as experiencers-of-lives should be acknowledged and increasingly worked into conservation plans and actions.

Many noted issues of conservation import, such as nativity and rarity, are rife with ambiguity and uncertainty. The problem of uncertainty, while a thorny aspect of science in general, is especially a part of conservation biology given its crisis-averting nature. Conservation biologists, according to disciplinary founder Michael Soule, "must act before knowing all the facts" (1985: 727). Not "are permitted to act" or "should act," "must act" prior to a likely outcome. Since the crises conservationists seek to address are plausible but presumably not necessary future events (i.e., extinctions) that come to be through a complex array of causes, influenced by numerous individual and corporate wills, in spite of the best modeling efforts we cannot know every outcome in advance. Even if we could know what the outcome would be for every organismic group in advance given certain conditions, it would be a giant leap to cause such conditions to obtain. To believe that this sort of control is possible is to view the world as a rather inert laboratory where humans are intentionally and methodically pulling all the strings for the good of its living inhabitants. This is clearly not the case, however, and one way of acknowledging our lack of perfect control, without giving up, is to build the capacity to address, both proactively and reactively, the variety of harms that can befall individual lives. This is a potential major point of agreement between wildlife rehab and conservation.

Are the values of conservation and wildlife rehab alike enough, on balance, for the practices to be considered complementary? Although we have seen that differences exist, these differences are not so extreme as to prevent complementarity. If we revisit the coin analogy and hold the large-scale preservationist tendency of conventional conservation to be one side of the coin, with the small scale restorationist tendency of rehab the other side of the coin, we have a near perfect example of a complementary relationship. What is more, we may posit an even closer relationship between rehab and conservation, where rehab is considered a subtype of conservation, a small-scale conservation practice of the sort that is necessary when large-scale preservation has already been achieved or is no longer plausible. Although debates will doubtless continue over the issues of nativity and rarity, we have at least explored potential ways of approaching these problems that may render them more soluble. One way is to consider the temporal vagueness and ecological uncertainty of the nonnative designation, and another is to think more deeply about the problem that data deficiency currently poses for conservation action. Those fixated only on species with current endangered status, moreover, would do well to consider the passenger pigeon (*Ectopistes migratorius*), which went from superabundance to extinction in the span of a human lifetime. Rehab and conservation, while not without tensions, are amicable overall.

As a final point on the relationship between conservation and rehab, consider an analogy between Plato and Aristotle, as depicted in Raphael's famous painting, *The School of Athens*. In it, Plato and his student Aristotle are walking together; Plato points upward, while Aristotle gestures downward. These different gestures can be interpreted as indicating Plato's emphasis on ideas, and Aristotle's emphasis on the empirically observable world. The traditional conservationists may be likened to the Platonists, theoretically pointing to heaven and striving toward the preservation of an ideal state of nature. They are the idealists, attempting to freeze a portion of the world at a particular time slice. Rehabbers may be likened to Aristotelians, pointing downward and seeing the necessity of dealing with the imperfection of reality to accomplish hard-won victories on the ground. Although differing in emphasis and with certain disagreements as to metaphysics and methods, there is much agreement between the two schools as to what is being served. Plato and Aristotle served truth generally, while both conservationists and rehabbers serve life in its diversity of natural forms. The "Platonists" and "Aristotelians" of wildlife may never come to agree on every detail, but will continue to travel together, learn from one another, and, I think, progress on a generally convergent path.

I have tried to help make such a convergence possible by bringing these formerly separate disciplines into dialogue. The way I have done this is by attempting to break down linguistic and conceptual barriers, thereby making what was formerly incommensurable, commensurable. Although conservation biology and wildlife rehab have been largely described in the same language, the worlds of meaning each has occupied have been theoretically and practically segregated. Hopefully, this segregation has been challenged enough to allow for a positive change.

I would like to close with a striking quote I recently came across by conservationist N. V. K. Ashraf:

> We as conservationists are caught in a paradigm shift from the past to the present and the future. That is, the shift from traditional prescriptions of leaving wildlife unattended to the present concept of managing wildlife for conservation and the need for an interactive management in the near future to maintain viable populations. With most of our wildlife living in pockets of few individuals with virtually little or no prospect of movement outside, wildlife translocations and rehabilitation will be increasingly employed as management tools to maintain population viability and genetic diversity. For many of these populations, the old model of wildlife 'preservation' and even the contemporary prescription of 'conservation'—wherein there is a hidden component of 'sustainable utilization'—is becoming less convincing and irrelevant to India. Wildlife rehabilitation as a conservation tool and a welfare imperative is part of a new paradigm in wildlife conservation that is slowly coming into being. (Quoted in Menon et al. 2005: 17)

This statement is not only relevant to India but should be seen as an indicator of the kind and magnitude of change in conservation and wildlife management philosophy that is beginning to occur around the world. It just so happens that the paradigm shift Ashraf has outlined above is both a moral improvement and increasingly practically necessary. Setting aside large tracts of habitat is something that North America, relatively speaking, has done well. But now is the time to begin gaining humility and practicing individual animal conservation.

References

Allen, T. F. H., and T. W. Hoekstra. 1992. *Toward a Unified Ecology*. New York: Columbia University Press.

Crook, R. J. 2021. Behavioral and Neurophysiological Evidence Suggests Affective Pain Experience In Octopus. *Iscience* 24:3.

Erb, J., and C. Humpal. 2020. *Minnesota Wolf Population Update 2020*. St. Paul: MN DNR.

Heister, A. 2022. *Beyond the North American Model of Wildlife Conservation: From Lethal to Compassionate Conservation*. Cham: Palgrave Macmillan.

Lynn, W. S., L. Baker, W. T. Borrie, A. P. A. Cardilini, S. M. Alexander, S. Coghlan, P. Cryer, G. T. Bonsen, T. T. Derham, O. Keynan, C. M. Reed, S. Riley, E. A. Ryan, F. J. Santiago-Avila, K. Walker, A. E. Weiss, and N. Xenakis. 2025. Compassionate Conservation in Practice: A Values-Driven, Interdisciplinary, Pluralistic, and Deliberative Community. *Biological Conservation* 303:111002.

Menon, V., N. V. K. Ashraf, P. Panda, and K. Mainkar, eds. 2005. *Back to the Wild: Studies in Wildlife Rehabilitation, Conservation Reference Series No. 2*. New Delhi: Wildlife Trust of India.

Sneddon, L. U. 2020. Can Fish Experience Pain? In *The Welfare of Fish, Animal Welfare 20*, ed. T. Kristiansen, A. Fernö, M. Pavlidis, and H. van de Vis. Cham: Springer.

Soule, M. E. 1985. What Is Conservation Biology? *BioScience* 35 (11): 727–734.

Wallach, A. D., C. Batavia, M. Bekoff, S. Alexander, L. Baker, D. Ben-Ami, L. Boronyak, A. P. A. Cardilin, Y. Carmel, D. Celermajer, S. Coghlan, Y. Dahdal, J. J. Gomez, G. Kaplan, O. Keynan, A. Khalilieh, H. Kopnina, W. S. Lynn, Y. Narayanan, S. Riley, F. J. Santiago-Ávila, E. Yanco, M. A. Zemanova, and D. Ramp. 2020. Recognizing Animal Personhood in Compassionate Conservation. *Conservation Biology* 34:1097–1106.

Appendix

Notes from Season One

1.

My first experience volunteering at the WRC was characterized by feelings and emotions of stress, anxiety, hurriedness, and helplessness, along with joy, gratitude, excitement, and even amusement. The more negative of these emotions, I surmise, resulted from my complete lack of experience in the field of wildlife rehabilitation, which was greeted by a barrage of new sensory data upon entering the avian nursery (my assigned volunteering area. Other areas people volunteer in include the Mammal Nursery and the Waterfowl Nursery. In addition, interns can work as Visiting Veterinary Students, Veterinary Technicians, or students of other wildlife- or biology-related disciplines.). These sensations, including sights (of volunteers, interns, veterinary staff, myriad signs and written instructions, and scores of birds of different ages and species), sounds (of verbal instructions given amid a cacophony of chirps, caws, shrieks, cheeps, and squawks), smells (of a mixture of bird poop, fish, and dog food), and feelings (of a fledgling blue jay [*Cyanocitta cristata*] claw, of beaks that close on my finger but do not pinch or hurt it), were immediately quite overwhelming.

L. B. DeVaney, *Animal Conservation Ethics and the Population Problem*, The Palgrave Macmillan Animal Ethics Series,
https://doi.org/10.1007/978-3-031-91891-9

The WRC is essentially a hospital for wild animals, and despite their name, hospitals are often not experienced as warm, welcoming places. This is unquestionably true for patients, but I have begun to see how such a chaotic, frenzied environment can be unsettling for workers, as well. On this cursory glance, this chaos seems, on the other hand, to be relatively controlled by a group of focused and fastidious human workers, and one of them takes me "under her wing" (an appropriate metaphor for such a place) and tells and shows me what must be done, and done repeatedly, during each weekly, four-hour shift. I did not expect to have to deal with numbers so much, qualitative social scientist that I am. I have hope that the overwhelming nature of my volunteering experiences will be gradually replaced by the same calm yet enthusiastic focus I discern among those experienced in this worthy, and relatively hidden, pursuit.

2.

Last night I substituted for a volunteer, per her email request (sent out to a large list of email addresses). Using this list, I was also able to find someone willing to sub for me this Saturday, during my brother's high school graduation party. Last night was only my second volunteer shift, and my first during this time (4:30–9:00 PM; my normal shift is 12:30–4:30 PM). Although at first glance it may seem a bit unfair to tack on the extra half hour for evening volunteers, after last night I can understand why this is the case. Some of the people in the avian nursery, who I assume were interns and/or paid staff, were discussing the possibility of staying until 11:30. They seemed to be rather serious.

The shift was filled with mostly new (human—I recognized some of the birds from my previous shift) faces, as well as new personalities. I was a bit surprised with the level of talkativeness of the workers in this shift, compared with my first shift. At orientation, we had been encouraged to not talk to the birds, and I assumed that this same principle would be in effect when talking around them, as well. This suggests another comment made by the orientation leader, to the effect that we should use technological devices as little as possible while at the WRC. She added that a past patient, a blue jay, had learned to mimic the sound of a telephone, and was subsequently euthanized. These statements brought up some nagging questions: Was it necessary to kill the bird? Would this mimicry have truly proven survival in the wild impossible, or even improbable? And wasn't the entire facility bursting at the seams, as it were, with technology? How are the medical devices and techniques being used impacting the patients? This last question was increased in salience, in my mind, by reading parts of Jeff Lederman's book *Cries of the Wild* (1997), in which the frequent use of homeopathic and naturopathic remedies in his rehab practice is discussed. Finally, it is encouraging to me to know that some men, such as Lederman, do in fact practice wildlife rehab; so far, all the other workers in the avian nursery have been women.

3.

There were a few noteworthy occurrences during my first shift in 10 days. Firstly, this was the first shift I have worked with a male volunteer. Secondly, this was my first shift not on incubators. I was on baskets for the first time, which are a sort of midway point between incubators and cages. Cages, which I spent some time on today (as well as my first shift), do not seem overly challenging; the most difficult parts for me are performing mental calculations of the appropriate amounts of Mazuri (a paste that is fed to most species of birds, about the consistency and color of peanut butter) to feed each of the up to seven birds in each cage, then making sure they are all actually fed. Baskets, on the other hand, are a different story. Whereas one can usually insert the food-filled syringe between the bars of the cage and into hungry mouths without being forced to open the cage door and risk escape, this convenience is less often the case with baskets. Often, one is forced to open the front of the basket to get to the bird (the back is held together with what looks like an industrial-strength paper clip).

Some birds at this stage are already so good at escaping, however, that they will slip through what seem impossibly narrow openings, which usually necessitates catching them with a net, or bare hands if they are less agile. At least a half-dozen birds escaped from baskets I was tending today, which was frustrating, a bit embarrassing, and potentially injurious to the already-stressed birds. Although notes are often posted on the clipboards that accompany each basket, with helpful messages such as "escape artist!" written on them, it remains difficult to predict an individual bird's behavior until I have tried to feed it at least once. Even after this initial exposure to a bird's tendencies, however, some are just too good (or bad, as the case may be). For instance, a particular common grackle (*Quiscalus quiscula*) whose care I was charged with today escaped at least four times in a single shift, between me and another volunteer. This necessitated some fancy net work on my part, which I was not exactly prepared to deliver. Usually, such escapism is an indication that this animal might be ready for a cage. Such a prompt transfer is not always possible, however, as space is often at a premium at the WRC.

4.

I have been volunteering for nearly a month now. I wish I was honestly able to say that the time has gone by quickly, but I cannot. Although I find the work fulfilling, and believe it to be important, it is hard work, marked by seemingly endless repetition and effort. This is a job in which it is necessary to constantly "shift gears" (or at least be ready to do so at a moment's notice)—not something I am particularly adept at. One is forced to adapt continuously to different species and their differing nutritional requirements, differing numbers of birds in each basket, the fact that some will take food willingly (even perching on the syringe, in the case of some very aggressive house sparrows [*Passer domesticus*]) while others will require the difficult and frustrating effort of force feeding, the ever-present threat of escape, etc. On the last point: I am heartened by the greatly reduced number of occurrences of escape that I directly contributed to last session (probably about half compared with last time). What this session lacked in number of escapees, however, was made up for by skill in remaining escaped. For example, one new admit—I believe it was a house finch (*Haemorhous mexicanus*)— had an astonishing ability to use every crack and crevice to avoid capture. I eventually found it in what had been a flat, closed fast food bag. I proceeded to grasp the bag by its top and carry my apparent Mcfinch sandwich back to its basket (or perhaps the bag only contained finch fries? I don't recall).

5.

This week I was solely on cages for the first time. Unfortunately for me, cages are a perfectionist's nightmare. The birds in the cages are at a bit of a transitional stage, somewhere between the confined dependence of baskets and moderate independence of the flight room. Before birds can be sent to the flight room, they must be demonstrably able to feed without direct assistance (aside, of course, from providing them with dishes of water and food). Since there can be up to 10 birds in a cage, not all of which necessarily came from the same nest (but which are of the same species), it is inevitable that some will get closer to self-sufficiency earlier than others. Because of this developmental gap, and because all the occupants of a cage are generally sent to the flight room together, some will also go to the flight room developmentally further along than others.

The reasons, then, that cages are a perfectionist's nightmare, are as follows: (1) Cages often have five to 10 occupants, all of whom look similar and most of whom move around frequently and vigorously; (2) Many cages are split rather evenly between those who will regularly voluntarily eat from a syringe and those who refuse. This is largely because of the developmental gap, as some are at the stage where they prefer to eat entirely on their own; moreover, this independence is a good trait, since it will be necessary for survival in the wild. I have found that it is difficult for me, however, to ignore the instinct to make sure that not only do all birds eat from the syringe, but that they all eat equal amounts. I have also found that this ideal consistently exceeds my abilities to both keep track of each individual and mentally perform precise fractional calculations concerning the volume of Mazuri per individual. If I had such problems with feeding in incubators or baskets, I would most likely force feed each bird (unless the bird's crop appeared full); however, at this stage it is necessary to have a bit of faith that the reluctant individual is consuming enough on their own. Letting go is indeed hard to do.

6.

I have now been volunteering at the WRC for over a month. I am very grateful to report that I have yet to find a dead bird during one of my shifts; however, I know from overhearing the conversations of others in the avian nursery that this happens from time to time. This could be mere coincidence, but I feel that the average health status of birds in the nursery is improving. Although I still see dehydrated (marked by visible lines on the abdomen and papery skin) and obviously ill birds, my perception is that they are, overall, healthier than when I began mid-June. This might be partially due to what I at least hope is an increased proficiency on my part in interpreting status and administering nutrition and hydration; or, given its unscientific, impressionistic basis, it could be nothing more than wishful thinking. In any case, I am beginning to feel more comfortable and confident in my ability, although I still find myself constantly glancing back at the feeding guide charts when I'm working on incubators (which I was this week). There are two kinds of these charts posted on the wall of the nursery for birds in incubators: one that gives appropriate food volume range per feeding for common species, and one that gives this range for progressively larger, to-scale drawings of baby birds (each basket and cage has its own sheet indicating appropriate food amount per patient). I am especially thankful for the drawings, as there are many species and even genera I have already come across which are not represented on the name-list. Although I realize it is not intended to be an exhaustive list, it could perhaps be expanded (e.g., I have not seen swallows on this list, which seem to be very common at the WRC). The diversity of birds has been a bit surprising to me, and I have seen a few young birds already that were marked as "unidentified" (regarding species). Perhaps the strangest bird I've fed thus far—and one that I certainly would not have been able to identify—was a common nighthawk (*Chordeiles minor*) last week. This bird, I learned, is not a real hawk at all, but is a species of nightjar. It is a plump bird with a capacious mouth, active at dawn and dusk, which lays its eggs on bare ground. Although I am of course sorry, on one level, that such a place as the WRC is necessary at all, on another level I am happy that such a venue exists where one can encounter and help such strange and wonderful creatures.

7.

I was saddened to hear, at the beginning of this week's volunteering session, that a nighthawk had recently died. Nobody was sure what the cause was. As I shared in my last entry, I fed a nighthawk last week. I immediately began to question whether I had contributed to its demise. Given my complete lack of familiarity with this species, perhaps I had done something very wrong in handling and/or feeding it. I was somewhat relieved to find, upon examining the cages, that a cage with a "common nighthawk" tag on it was empty (presumably meaning that this was the one that died). I quickly checked to confirm that the one I had worked with was still in the incubator, alive (it was). So, given that I do not recall working with any nighthawks in any cages, I think I can safely assume that I didn't directly contribute to this bird's death. Fine. But why do I care so much that I am able to prove that I had nothing to do with it? Does it really matter, for the bird, whether I, someone else, or some completely impersonal force caused its untimely death? I doubt it. Here, I believe I am feeling the tension between what matters more, from a moral standpoint: The fact that something bad happened or the potential fact that I contributed to this bad thing happening. Should I be more concerned about having "dirty hands" as a person, or about the consequences that occurred and that some other person or force (or possibly a great many people or forces) caused? Or should I not be concerned at all? After all, what happened happened. Or is this simply a false dichotomy? I have no answer at this point; but I will not be surprised if this topic comes up again.

8.

This week was a bit overstaffed, which I guess is better than the opposite, at least. The multitude of volunteers and interns left me, however, feeling as though my presence in the room was more obstacle than aid. I managed to feed two apparently extremely hungry rock doves before realizing that there were two other people on cages— which meant that there really was nothing for me to do in the avian nursery. I peeked into the waterfowl nursery to see if I could be of assistance, only to find a room devoid of humans but full of ducks (I was told, puzzlingly, that they don't get going until the evening—I'll ask for clarification on that later). I was on the verge of giving up the quest for work when I found a vet tech and intern and asked them what to do. I was told that, if nothing else, I could do dishes or laundry. I chose to do the latter, and the intern quickly showed me the laundry room procedure.

The laundry room contained three conventional washers and one commercial dryer; the dryer had a plaque on it indicating that it had been donated by a woman in celebration of her 25th anniversary of graduation from college. Perhaps the most notable departure from normal laundry-washing was the necessary step of shaking out each individual item (most of which were towels, washcloths, or something that looked like a small mattress pad), so that no animals that might have somehow found their way into the laundry would get into the wash as well. I recall from the orientation session I attended that, unfortunately, some animals had been washed and/or dried to death in the past.

There was a plaque on the wall stating that the wearing of gloves and a mask is mandatory. I soon found there to be good reason for this, as many of the laundry items were covered in feces (among other things). The stench was astonishingly bad; I thought at one point that I might vomit. Some of the laundry had apparently been there for a while, as there were tiny maggots wriggling around in the feces. I could understand, now, why there were baskets of dirty laundry not only in the laundry room, but lining half of the adjoining hallway. I was happy, at least, to get over half of it done during my shift—lord knows when the next brave soul would tackle the remainder. Lastly, I was reminded by a single piece of flypaper hanging from the doorframe that the compassion of even the most compassionate individuals and organizations is not boundless.

9.

Sunday, July 31, I found a tiny, all-but-naked baby bird of unknown species in the alley behind my apartment in Falcon Heights, Minnesota. I spotted it as I was at the point of near exhaustion, after a long walk, approaching my apartment from a direction I would not normally come from, but had decided to tack onto the end of my walk to add some time. I immediately thought the worst, as I saw it lying there, motionless, helpless, and probably severely injured, on the rough asphalt. I picked it up and was glad to see that it was still alive. Although I scanned the area, including the eaves of nearby garages, I could not find any nests; so, a bit perplexed as to its origins, I brought it in, and my friend, Miriam, proceeded to place it in a small plastic container with a paper towel in it. (On closer examination, it turns out that the bird probably fell from a nest in a plastic box on the electrical wire high above the alley.) She drove it to the WRC, only to find that it was closed. So, it looked like the poor little bird was stuck with us for the night.

Although we both wanted to give it water and food, it is generally recommended that baby birds not be offered food or water until they can be examined by a vet (unless, of course, for some reason they must be with you for a long time before this is possible). Since I figured that the probability of serious injury was high, we forced ourselves to abide by these guidelines and hope for the best in the morning. The bird appeared to be shivering, so we placed it (in its plastic container) in an empty mouse cage (so that it wouldn't be bothered by our cat [*Felis catus*], Squiggy, or dog [*Canis lupus familiaris*], Tonky), on top of the heating pad. After waiting a while and checking on it, it was still shivering, so I made it a nest and "blanket" with toilet paper, and loosely placed a lid over the top of its container so that it would retain more heat (while still allowing air in to breathe). "Petey," as we had named it, would have to tough it out for the night.

The next morning, I expected the worst but was delighted when Miriam rushed in and told me, "It's chirping!" Sure enough, Petey was emitting astonishingly loud chirps, undoubtedly out of hunger and thirst. Miriam proceeded to leave for class, while I waited for the WRC to open (at 9 AM). Unfortunately, my car had recently stopped running, and I had yet to fix (or junk) it. I thought that this wouldn't be a problem, however, as Miriam had taken her bike to class and left the car with me. I hadn't anticipated that she would take her key with her, and I would be unable to find my key—which, of course, was exactly what happened. After nearly turning the apartment upside-down in panic, I relegated myself to the fact that I wasn't going to find the key. With no other visible choice but waiting for the bird to get more dehydrated and emaciated, I stuffed his little container in my pannier bag and headed off on my bike. Luckily, it is only about 5.6 kilometers from my apartment to the WRC; however, I knew it would be a bumpy ride, and was a bit concerned about this. As I rolled into the WRC parking lot and locked my bike to a bench, dripping with sweat and rain, I was again relieved to hear the bird's vigorous chirps. I dropped Petey off with the receptionist at the front desk and headed back home, happy that I could at least check on his status during my shift that Saturday.

Five days passed, and I arrived at the WRC with only one concern: To check on Petey. I entered the avian nursery and searched the incubators thoroughly, but was dismayed that I couldn't find anyone there that remotely resembled Petey. Steeling myself for the bitter news, I proceeded to ask the receptionist his status. To my astonishment, the case report associated with my name on August 1 stated that the bird was medically healthy, and I rushed back to see if I could find the case number in an incubator. I managed to find it and was further awed by the avian bearer of this case number, apparently healthy and…fully feathered! I asked an intern if this was possible, and she replied that it was, indeed. I guess the every-15-minute feedings go to good use, after all. Petey is a house sparrow—a nonnative species—but I could care less. I hope to find him hale and healthy next week, as well.

10.

Tuesday, August 9, while house-sitting for my in-laws in nearby New Brighton, upon looking out the window I was a bit surprised to see a duck on the front lawn. I was not shocked, however, as I knew there to be a small pond just across an abandoned road adjacent to their property. Perhaps the female mallard (*Anas platyrhynchos*) had decided to stop and take a rest before the final leg of her journey? This possibility began to look less probable, however, as time passed and I didn't notice any significant movement. Moreover, the duck appeared to have flies swarming around her, which she was weakly biting at. Something was wrong. Miriam noticed her, as well, and we slowly moved in for a closer look. The duck's back seemed to be wounded, and one of her legs looked broken. However, she evidently still had the power of mobility, as she took a short flight into the neighbors' yard when we tried to examine her more closely. I found some work gloves in the garage, and set off to capture her, while Miriam followed close behind with a cardboard box.

The duck promptly scuttled under a very thick cedar tree with branches to the ground, which made it difficult to get close. I managed to get one hand on her, but barely, and she decided to take a longer flight this time—apparently into the woods. Although one of her legs was badly injured, she appeared to be able to fly well, so—as we were already running late for a dinner date with some friends—we decided not to pursue her any further, lest this turn into a wild duck chase. I was reluctant to let the duck go, though, as I knew that what we had just witnessed could have constituted a final expression of her will to live, were she not to receive any medical aid. I took one last glance at the area in which I thought she had landed, and, to my surprise, she was on the edge of the woods, still in the yard, and hemmed in by a fence. I walked slowly toward her, resolved to not let her get away this time. She tried to work her way under the fence, but to no avail, and this time I got a firm hold of her. She protested weakly, but did not try to bite or peck me. I carried her to the car and placed her in a box with an old shirt. As Miriam drove us to the WRC, she decided her name was Sarah. For some reason, the name seemed to fit. Dinner would have to wait tonight.

11.

The season is beginning to wind down. It has been communicated to the volunteers, via email, that the avian nursery will officially close for the season at some point in late September. While it will be nice to have my Saturdays free again, I will certainly miss the avian nursery. Despite the hard work involved in the seemingly endless iterations of feedings, volunteering at the WRC has been one of the most fulfilling experiences of my life thus far. As I was on vacation last week, I found myself looking at the world through a slightly different "lens" than I had been on previous leisurely trips I'd taken. For example, I seem to have developed a tendency to examine "wildlife" (if the animal residents of New York City can accurately be referred to as such) more closely than before. Whereas previously I don't recall having studied the bodily condition of birds I am in proximity with, I noticed myself attending to the feet of pigeons, many of which appeared to be missing toes or otherwise crippled.

One rather disturbing incident I feel I should relate, but that I didn't think to include in the previous post, occurred in New Haven. Miriam, her brother Carl, and I were walking to a coffee shop. As we were passing Wooster Square, I noticed a lifeless-looking bird on the sidewalk. Before I was able to say anything about it, Carl absentmindedly stepped directly on it. If it wasn't dead before he stepped on it, it certainly was after. I don't relate this with hope that it reflects poorly on Carl—it was I, after all, who, as an adolescent boy, intentionally and wantonly shot, with an air rifle, members of many of the species that I'm currently trying to help. I think this recent experience, though, illustrates the ease with which humans, both metaphorically and literally, can tread upon nature without even being aware of it. According to the National Wildlife Rehabilitators Association website, "over 75% of the animals cared for are affected in some manner by human activities." I think this could be stated more strongly without being inaccurate. While more information about the harmful effects of certain human behaviors and technologies is helpful, and indeed crucial, for humans to live harmoniously with wildlife, it is indeed possible to cause harm completely unintentionally, participating in the most mundane and assumedly harmless of activities. Perhaps the storied Jain practice of sweeping the path in front of one as one walks is not merely obsessive religious perfectionism, as it may immediately appear to many Westerners, but an acknowledgment of the real possibility of harm that accompanies every step.

12.

Today ended up being a bit depressing, but did not start out that way. I commenced my shift with an expectation that the day would be manageable and enjoyable, given the three containers occupying the incubators. This was by far the smallest number of patients I had yet seen, not only in the incubators, but the whole of the avian nursery. Many of the baskets were gone, as well; only the cages seemed relatively full, with a majority of these tagged "do not feed" (meaning do not assist them in feeding by hand, hemostat, or syringe, but make sure the dishes in the cages have food and water in them at all times) as the residents were being prepared for independence prior to being sent to the flight room. For the first time I recall, I was able to handle all of the incubators alone, with time to spare at the end of every 15-minute round of feedings. During this spare time, I checked on the cages on my side—giving them fresh water in clean dishes, and providing them with grubs, mealworms (*Tenebrio molitor*), or fruit (after initially mistakenly feeding a cage of cedar waxwings [*Bombycilla cedrorum*] grubs with a hemostat before noticing the small sheets of green paper warning against assisted feeding on this and three of the other six cages on my side of the room).

All seemed to be going well, save for a goldfinch who refused to gape and had to be force fed each round. At some point mid-session, I noticed a chimney swift (*Chaetura pelagica*) that an intern had presumably placed in one of the incubators. On its container was a note reading, "Do not feed – drying off." Assuming that it had simply taken a bath in its water dish and gotten cold, I left it alone for an hour or so. When I finally checked on it, it was obviously in bad shape—propped up against the side of the container and very lethargic. I noticed that one of its wings appeared longer than the other. I showed it to the interns, who agreed that it didn't look good. One of them said something about a vet check (which I don't think occurred). After asking whether I should give it anything, and receiving a negative reply, I put it in the warmer of the two incubators after confirming that this was okay. Shortly after this, an intern came over to inquire about its whereabouts, and to my surprise, it was no longer in the incubator! It turns out that another volunteer had, unbeknownst to me, removed it and was either force feeding it or giving it LRS (a rehydrating solution) —neither of which seemed a bad idea to me. The intern told the volunteer to put it back in the incubator, which she did. The next time I checked on it, it was moving a bit, and both an intern and I decided it looked a little better. The next time I checked on it after that, it was dead—stiff, eyes open, feet clenched. This was surprising, dismaying, and a bit disturbing, as it was the first death I had witnessed at the WRC. An intern took the dead swift out of the nursery, and was reminded by another intern as she left to "take off the leg band."

This experience reminded me to check on the statuses of Petey and Sarah, which I hadn't done for several weeks. Petey grew up well and had been released. Sarah, however, had to be euthanized, as I was told that her elbow joints were so badly luxated (dislocated) that recovery was not possible. This brought the success rate of animals I'd participated in rescuing and bringing to the WRC to 50%—one eastern gray squirrel and one house sparrow, both rehabilitated and released; one eastern cottontail, died in cage; one mallard, euthanized. Not as good as I'd hoped for, but not atypical, unfortunately.

13.

When I arrived, the Avian Nursery Coordinator was the only human in the room; apparently, come September, the volunteers and interns begin to thin out. I was a bit worried, as I had broken the dress code on two counts by wearing sandals and shorts (I had gone to the WRC to volunteer from a volleyball game my sister was playing in and had neglected to dress appropriately), but the coordinator was happy to just have another pair of hands to help. I was essentially told that I could work on whatever I wanted, and, as has been the norm lately, I chose incubators, since their occupants must be fed the most frequently, and don't have their own supply of food and water for self-feeding.

Surprisingly, there were roughly twice as many patients in the incubators this week as last, the majority of which were American goldfinches (*Carduelis tristis*). Goldfinches, I was told, tend to nest late in the year, and my recent experiences at the WRC support this. Between the incubators and baskets, there were literally dozens of them this week. Luckily, another volunteer arrived, albeit somewhat late, ensuring that we at least had one worker for every type of enclosure. Although it is normal protocol to pre-fill syringes with the special grain-based finch food mixture and keep them in the refrigerator (as it spoils relatively quickly), the volunteer on baskets simply kept all of the filled syringes with her, as she was using them as fast as she could fill them. It seemed I could expect fewer breaks this week than last, between feeding and filling syringes.

I greatly enjoy working with goldfinches, as they are cute, trusting (it's often a challenge to keep them off my hand), and usually eager eaters. They tend to be such good eaters, in fact, that one must watch their crops closely; they are tiny birds, and easily overfed. It has been rare, in my experience, for a goldfinch to escape. This week, however, the only occupant of Incubator C took an unexpected flight straight into the door. Thankfully, it didn't have much room to accelerate before contact and seemed unharmed from the incident—which prompted discussion of possible ways of preventing such occurrences in the future, for example, by stretching a net or screen of some sort across the doorway, as had already been done across the large window.

Notes from Season Two

1.

This season of rehabbing has started much earlier than the last, and differently. While last year was spent focused on the avian nursery, this year will be split between the mammal nursery and the waterfowl nursery. I have been told that, because of the heavy load of squirrels the WRC receives and the two distinct breeding seasons of squirrels that occur in spring and late summer/early autumn, the mammal nursery effectively shuts down during the mid-summer months. During this time, I plan to volunteer in the waterfowl nursery, thereby obtaining experience of the entire WRC nursery trilogy in just two seasons. Although I had wondered why the mammal nursery is inactive between the two breeding seasons (don't other mammals need care then?), my first day today cleared up some of the confusion.

The entire mammal nursery consists of squirrels (aside from the human volunteers and interns, of course): eastern gray squirrels, American red squirrels (*Tamiasciurus hudsonicus*), fox squirrels (*Sciurus niger*), and flying squirrels (genus *Glaucomys*). Partially to keep predators and potential prey separate, and partially because of rabies concerns with many other mammal species, other mammals are kept in a separate, as-yet-unseen part of the facility. All the squirrels I attended to today were gray squirrels, by far the most numerous of the four in this area. "Gray," much like "black" in black bear, is, somewhat confusingly, part of a species name but only signifies the most common coat color. Today alone, in a rather small sample of admits, I encountered everything from white (albeit albino) to charcoal colorations.

Although the mammal nursery is only a door over from the avian nursery, it is a different world inhabited by very different beings. I will proceed to list some specific differences I have noticed. These beings, I have learned already, bite, scratch, and climb more than the others, but escape less and, on average, are much easier to feed. More like human babies, they drink formula instead of consuming foul-smelling tan goop. Instead of dozens of feedings per day, squirrels are fed four times per day at most. Unlike the room next door, there are no incubators (that I noticed, at least). There are sign-in and sign-out procedures, as well as more detailed forms to fill out for each individual feeding. Instead of leg bands, these critters are marked for identification with different colors of nail polish on their ears. They must be weighed before every feeding. There is a special formula to feed them if they have diarrhea. Instead of an indoor flight room, squirrels that have graduated from the nursery are sent to a large outdoor cage on the adjacent deck, replete with lots of enriching leaves, pinecones, and other natural objects. Finally, each basket or cage has a tiny fleece hammock for the squirrels to lounge in, which they seem to thoroughly enjoy.

A general, initial impression of the mammal nursery is that, because of the far fewer feedings necessary per patient per session, more time can be devoted to assessing the condition of, and caring for, each individual organism if there is a full volunteer load. I was a bit surprised today to find four to five volunteers (including myself) at such an early shift (8 AM–noon), and on only the day after the opening of the nursery for the season. To see this kind of dedication is encouraging, and exactly what is needed if wild lives are to be spared in the increasingly hostile battle between wild animals on the one hand, and humans and their technologies on the other. It must be admitted, however, that such an analogy of war is only partially appropriate. To invoke war is to imply that each side can defend itself reasonably well, and the fact is, for wild animals there is no adequate defense for cars, guns, traps, and the innumerable other unnatural obstacles they face. Although such unfairness is felt at the level of the individual, it also wreaks havoc with the entire process of natural selection. Thanks to us, evolution is now more of a crapshoot than a competition. What wild animals are now experiencing in industrialized nations is not, however, unlike war in that it is unvaryingly worse. Neither all humans, nor all technologies, are "the enemy" to wildlife, and the practice of wildlife rehabilitation proves this.

2.

Today was my first Tuesday shift, and somewhat more stressful than my initial Thursday shift. Upon arrival, as I was about to lock up my bicycle as I had always done in the past, I was told by a WRC employee that I should bring it inside—presumably because of the danger of theft. So much for my idea of Roseville as the "perfect" suburb. I entered the building to find that there are different shift leaders on Tuesdays and Thursdays, who apparently have quite different organizational styles. This shift seemed more formal and rigorous in comparison with the last, with much procedural detail and constant reminders, usually followed by "Does that make sense?" Some of the difference I sensed could be due to personality; however, I noticed several more tangible differences, as well. I will proceed by listing and discussing some noteworthy distinctions that I have perceived between the two at this early stage.

Recordkeeping procedures were markedly different today, and possibly more organized. Rather than each volunteer taking a book in which to record weights of squirrels, amount fed, strength of formula, presence or absence of diarrhea, etc., individual files were removed from the three-ring binders pre-shift by coordinators. When all the duties pertaining to a particular basket or cage are completed and every relevant field filled in, the files are placed back in the binder (hopefully in order—the first time I attempted this, I put the file on the wrong side of the marker tab). Each file consists of from one to several individual sheets of color-coded paper stapled together, depending on how long the patient has been in the nursery. The color coding denotes the number of feedings designated to each patient per day, which is in turn determined by the weight of the squirrel—the lighter, the more feedings, and vice versa. When a patient surpasses a certain weight threshold, they "graduate" to fewer feedings (or self-feeding, if they are quite heavy).

It was strongly and clearly communicated that the feeder must be careful not to press the syringe so fast that it comes out the squirrel's nose, indicating aspiration, and to not push so slowly that a sucking sound is heard. Unfortunately, many of the syringes make it difficult to modulate the delivery of the formula precisely. We were reminded to wipe off the tip of the syringe before filling it to minimize the amount of formula on the squirrel's coat, and to clean the mouth area of the squirrel with a wet cotton ball or pad. Cotton balls are also used to stimulate defecation, if necessary, by rubbing them on the squirrel's underside. I was told to do this to check for diarrhea today, to "see what comes out" (nothing came out).

Today, I gave my first round of diluted formula to a recent admit. There are two basic kinds of formula: "eyes open" and "eyes closed." I am not sure yet exactly what the difference is but was told that one is stronger than the other. I have not yet fed, or even seen, any squirrels with closed eyes. The first thing a patient is fed when admitted is "LRS," a hydrating liquid (the same that is fed to dehydrated birds in the avian nursery). They are then gradually introduced to stronger and stronger formulations. The mixing of different formula strengths is accomplished by simply drawing the appropriate amount of LRS into the syringe, then filling the remainder with formula. I also fed an albino grey squirrel for the first time and witnessed a red squirrel in the nursery for the first time.

Regarding the different communication styles experienced between the two shift leaders: It is difficult for me to tell, at this point, which style is more effective (or even if one is likely to prove more effective than the other), as each seems to have its own merits. While no attention to detail can be too great in such a context, considering the incredible number of variables involved in care for each organism, enforcement of such protocol can cause stress for the volunteers and may be counterproductive to the goal of maximizing retention of experienced volunteers.

3.

Thursday shift two down, and I am optimistic about this being an enjoyable shift to work this season. A new volunteer, a retired high school teacher, joined the nursery today. In keeping with expectations deriving from my experience in the avian nursery last season, I am so far the only male volunteer. The coordinator of the mammal nursery, however, is a young man. The new volunteer seems quite witty and made a pun that was lost on me regarding the squirrels in basket C4 being "dynamite." I responded to the effect that I wasn't in pun mode, which may have merely been an excuse for my slow-wittedness. I don't think this was the only reason, though, as I consider it important to remain focused on the numerous tasks at hand during my shifts in the nursery.

Today I warned a volunteer about to feed and clean the cage of the so-called self-feeding squirrels about their tendency toward aggressiveness. She was wearing thin fabric gloves instead of thick leather ones and seemed blissfully unaware of the behavior she was about to encounter. It turns out it was a good thing I warned her, as this was her first time interacting with them and she was not expecting such aggressiveness. These grays are nearer to release than most of their nursery mates in plastic baskets (the same type of baskets used in the avian nursery, and which must be repaired with greater frequency [with duct tape] here because the squirrels tend to chew on them) and are kept in wooden cages.

I fed and cleaned cages for several particularly aggressive self-feeders for the first time last Tuesday. Upon approaching the cage, I was greeted with growling and was happy to be wearing thick gloves as I was bitten numerous times while trying to get hold of them. As many of these cages have four squirrels in them, often with different colors of nail polish on their ears that must have been mistakenly applied, it is extremely difficult to both identify and capture them for weighing, as well as keep any from escaping. Unfortunately, I had numerous escapees last Tuesday, with one particularly aggressive squirrel seeming to jump at my face from the head-height cage and, despite my attempts to catch it, fall to the floor. Not only is this a safety issue for the squirrels, but also for volunteers who might instinctively try to grab a loose squirrel with gloveless hands and get bitten. I will have to be more careful about preventing escapes in the future. Not all caged squirrels are aggressive, however. I could hardly keep one from benignly crawling out of its cage while I was trying to clean it, so I let it use me as a human "tree" while trying to avoid getting bitten by its more aggressive cage mates (I believe the docile squirrel was from a different litter than the others). I wonder how this aggressiveness or lack thereof will play into future survival prospects once released.

Today I witnessed a red squirrel feeding for the first time and was told that reds are generally more difficult to feed than grays. This seemed a bit ironic, however, because as I was being told this I was feeding a gray that would take only a fraction of a milliliter of formula at a time before taking a "break"; on its form was marked "VERY slow feeder." I had been told previously to press gently on the sides of a reluctant squirrel's mouth with my fingers to create suction, thereby helping the formula to flow from syringe to mouth in a steady stream, but I was unable to make this work well, so I resigned myself to feeding this very slow feeder a tiny bit at a time.

Unlike last Tuesday, we finished up about a half hour early today, so I was left with the task of deciding whether to leave with the others or stick around and find something to do. Knowing that I would have to miss several shifts this month, I settled upon the latter. I had a difficult time with this, as presently all the laundry machines were in use, and I was unable to find the director who I had thought to chat with if he was available. I was left wandering rather aimlessly around the center, feeling awkward and a bit out of place but hoping I would discover something useful. I ended up in the surgery area, a large open room with veterinarians and veterinary staff, operating tables, and instruments galore. As I entered, I overheard someone who I did not recognize as veterinary staff proclaiming loudly and vociferously how she was not in support of the "no-kill" designation, and how those who are must be uninformed. The veterinarian, who was working on what appeared to be two raccoon (*Procyon lotor*) kits, said that she agreed. I was curious about the context of this conversation, but as I was not intended to be a part of it, I did not ask. I did, however, ask the veterinarian if I could observe her. She replied that it would be better if I contacted the head veterinarian in advance about "shadowing" a vet and proceeded to write down an email address for me on a piece of paper. I took the hint and walked out to see if the laundry was done yet. As I reflected on this experience, I realized that what I had perceived as anesthesia may have in fact been the euthanasia of the raccoon kits. If the kits were indeed being euthanized, this would help explain the topic of conversation I walked into, as well as the vet's deflection of my request to observe. Although I cannot prove my suspicion, it is something to consider—as well as the question: If the kits were being killed, why?

4.

This session seemed uneventful. I have not been working with the self-feeders lately, and so have not had to deal with any escapes. A new volunteer joined us, and, as we finished early (in contrast with the last Tuesday shift), I finished up by washing baskets for the first time in the "cage wash" area. As the EnviroCare cleaning agent container was nearly empty, the new volunteer and I were led downstairs to where the refill barrel is kept and instructed in how to fill the dispenser. The barrel was nearly empty, and it seemed that every pump of the handle brought up only a small amount of liquid. We finally decided that three-quarters full was sufficient. We were also led into the general storage area, where the crew leader found that we were out of the favored squirrel food, Zupreem. I thought this odd, as it had been discovered only about a week and a half ago that we were out, and a new shipment had supposedly arrived since then. Perhaps the preference squirrels show for this primate food has not been exaggerated—ironically, it is reputedly much preferred to the fallback option, "rodent chow."

I began by stating that this session seemed uneventful. After writing this, however, I remembered that there was nonetheless something noteworthy about this session. Unfortunately, a gray squirrel, a self-feeder, was found dead in its cage during the morning session. Perhaps I did not interpret this as an event because, in a sense, the most noteworthy event of death had already occurred by the time it was discovered. Not much emotion was shown about this, overall, although one of the crew leaders (this session appears to be co-led) made a comment to the effect of "poor thing." The volunteers were instructed to be sure to write "DIC" (for died in cage) next to the relevant case number on the identifying card on the front of the cage or basket in which a death was discovered. Interested volunteers were then led into an area adjacent to the main veterinary examination room with several binders and records. The initial check-in form for the dead squirrel was found, and its death was recorded. The squirrel was then placed in a covered trash can. I asked sarcastically whether there was any ceremony and was told that there was not even a postmortem examination.

Sometime during the middle of the session, it was stated that many squirrels were ready to be released, and volunteers for releasing them were requested. I jumped at the opportunity, as I had not previously been involved in any releases. Although the rescue and release of animals are often the main foci of attention, they are no more important than the choice of release site and associated procedure. One other volunteer, as well as the crew leader who made the announcement, agreed to conduct the releases. We were told to release our respective groups in different areas—preferably with oak (genus *Quercus*) trees—so as not to overpopulate any one area. The volunteer stated that she would release her squirrels at Theodore Wirth Park in Minneapolis (Hennepin County), while the other planned to release hers at Lebanon Hills Regional Park in Dakota County. I said that I would release mine in Como Park, St Paul (Ramsey County)—a nice, county-per-carrier distribution. When I asked how many squirrels were in the plastic pet carrier I was handed, I was told “a lot.” I looked inside the carrier, and saw a wooden box containing the squirrels, with a hole cut in it large enough for the squirrels to fit through. I was told to tip the box up if any of the squirrels were reluctant to exit. I then brought the carrier to the one-way (only openable from the inside) gate at the front of the facility, and agreed to meet the crew leader outside with my car. I drove up, she brought out the carrier, and I placed it in the trunk of the car.

Although Como Park is on the way back to my apartment from the WRC, I decided to keep driving and pick Miriam up first, as I thought she would enjoy being a part of the experience, and that I might need some help. After picking her up, I realized that I did not know the best place to release the squirrels. I started mentally scanning all of the candidate areas in the park I could think of, and settled on a fairly open area, save for several oak trees, adjacent to a more densely wooded strip along Hamline Avenue. It was difficult for me to choose this place, however, as it seems that squirrels can thrive in nearly any environment with trees and a reliable food source. I wondered whether I should release them in a more secluded area, such as the Reservoir Woods, but was unable to think of a place there with many oak trees. I also wondered if anyone—such as the driver of the passing St Paul Parks vehicle—would see us, stop, and ask what we were doing, or even protest our actions. It turns out that nobody even seemed to notice, which I was quite grateful for.

I was a bit nervous about opening the cage, since I had no gloves, and expected a mad dash to exit the cramped quarters. Surprisingly, however, the grays did not leave immediately, or even seem to want to leave. We tried taking the wooden container out of the carrier, tipping it forward, holding it upside down, even shaking it, but the only result was the emergence of nine—yes, nine—bushy tails. I thought about trying to take the box apart but decided it would be better to just wait it out, hoping that the braver and more inquisitive of the inhabitants would make their way out given some time. This is just what happened, but slowly. One by one a squirrel would peek out, look around for a minute or so, and exit the box. Most of the exits, however, were more tentative than I expected, even hesitant and a bit confused. Two different squirrels seemed to immediately prefer the option of climbing my leg to that of climbing the closest tree, but each changed its mind and ultimately chose the tree. Despite waiting what seemed like an eternity, some would only come out with vigorous shaking of the box. The final squirrel absolutely refused to leave its cozy but now comfortably roomy "house," and I ended up wrapping my hand in a towel and pulling it out with much growling and shrieking ensuing.

The pet carrier contained many pieces of dry Zupreem and one almond, which I scattered about the lawn for the newly freed squirrels. I was a bit surprised at how small they were—not nearly as large as most "wild" grays I see in the area—and felt concerned for their ability to forage and detect potential harm. A few of the squirrels, including the last and most difficult release, seemed to want to stay close to Miriam and me, moving far out on the lower branches to get closer to us. At this point, I was suspicious of imprinting, and my suspicions were confirmed later that day. Miriam reported that she jogged by this area with Tonky that evening and was surprised to see two grays bounding toward her. She recognized one as the last, reluctant releasee, due to its reddish face and small size. Astonishingly, the pair came within centimeters of Tonky, apparently sniffing her while Miriam held Tonky's mouth closed. They then bounded away. It seems that they had followed Miriam, recognized Tonky as strange if not particularly a threat, then decided to leave. These signs both of imprinting and lack of recognition of a potential predator as such are discouraging and bring up questions about future methods for mitigation of this dangerous behavior. Although rehabilitators often go to great lengths to avoid imprinting in rare and endangered species, using species-specific puppets or even putting on full-body costumes resembling an adult of the patient's species, the lengths to which rehabilitators will be willing to go for common species such as gray squirrels even if solutions are found remains to be seen.

5.

I've been absent from the WRC for a few weeks now, having taken a trip out east with Miriam for her brother Carl's graduation (PhD in physics from Yale University). Upon opening the door to the squirrel nursery today, I heard the shift leader state that we could go home early today if we wanted to. It was immediately apparent why this was possible, as there was a mere fraction of the former patients present and a full volunteer crew. Apparently, the "down" season is well upon us—before I had even expected it to start.

There seems to currently be a greater ratio of red squirrels to gray squirrels in the nursery. I fed my first red today, and upon asking whether there was anything I should know about them, was told that "they never stop moving." This has proven to be the case, as the reds I worked with today seemed to prefer climbing circles around my torso to remaining still. I did not experience any reluctance to eat that I had previously been told characterizes reds, however, as my basket patient took the formula very readily. In addition to this feeding, I weighed two self-feeding reds today. Although I expected to be greeted with a wild ferocity even greater than that of the self-feeding grays (and, at least, a propensity for escape, due to the sign on the cage reading "escape artists"), I was surprised to find the reds quite docile. I had no trouble taking them out bare-handed – which I would not even try with caged grays, at this point—and found that the pair were only interested in escaping onto me. This observed behavior of reds and grays at the WRC is quite the opposite of what I expected based on observations of these species' interactions in the wild (using "wild" loosely here, of course, given that such interactions have been seen out of my apartment window).

Even with the "extras" like laundry, washing cages, etc., we finished about two hours early today. I offered to help in the avian nursery, but there were so many people in there already that I would only have gotten in the way. As a bit of an aside, I have been consistently observing surprising numbers of volunteers at the WRC—perhaps more than even such a distinguished organization can handle. Such power needs to be harnessed for wildlife! This reminds me of a quote from Schweitzer to the effect that he perceived there to be more reverence for life in the United States than in any other country. Part of me is very skeptical of this statement, especially considering the popularity of things like hunting, meat-eating, and industrial and corporate development in America. However—as ironic as this may be—I believe outdoor activities such as hunting and fishing to be potential gateways to a more robust and consistent respect for life. If it is indeed possible to channel or transform hunting zeal into a less harmful way of encountering nature, then Schweitzer may have been onto something big.

Given the large number of entries on the "ready for release" whiteboard, indicating animals that may be released by volunteers or staff after conferring with medical personnel, and the fact that I had driven, I thought it would be good to take the opportunity to release a different species of animal (the last being a cage of gray squirrels). Since these grays had been brought to the WRC as young, and had not established territories, they could be released anywhere with appropriate habitat. Animals brought in as adults, however, should be released as near to the place they were found as possible—and it seemed that most listed on the board were adults.

After rejecting several candidates because of lack of familiarity with their release area, I settled on a single eastern cottontail to be released near St Catherine University in St Paul. Recalling the lack of success I had with my first cottontail (also my first-ever attempt at a rescue), and things I had read about rabbits dying very easily from shock, I tried to take special care in the handling and transport of the soon-to-be releasee. It was handed to me very neatly packaged in a taped-up cardboard box, and I could scarcely believe that the box contained a rabbit. Judging from the size of the container, it was clearly not a large rabbit. I put the box on the floor of the car, so that it would not slide off the passenger seat, and turned the radio off to minimize noise.

I found the location surprisingly easily but was also surprised to find that the listed intersection was in a thoroughly residential area. If I was to follow directions, I would essentially have to release the rabbit in someone's front yard. I opened the box, gently lifted the small rabbit out, and placed it on the grass, then retreated to my nearby car to watch. The rabbit remained nearly motionless for a few minutes, its ears flattened against its body. I was surprised to see a gray squirrel amble across the street, seemingly curious about the "new" addition to the neighborhood. Could it have recognized the rabbit from times past? The squirrel hesitantly crept closer, until it was literally centimeters away from the rabbit, sniffing it. The rabbit now seemed alert—sitting up, erecting its ears, and looking back at the squirrel—but not threatened. The squirrel, satisfied, went on its way, while the rabbit hardly moved. After the passage of a couple more minutes, a man walking a large dog strolled by. Predictably, the dog saw the rabbit and showed interest. Thankfully, the rabbit quickly retreated into the back yard.

6.

Not surprisingly, another short shift. This one lasted less than two hours. I am surprised to still see new volunteers trickling in, although perhaps I shouldn't be, given I did not even start volunteering last year until mid-June. The leaders of both the Tuesday and Thursday shifts have now made it clear that we can (and perhaps should) alternate weeks for volunteering, given the "problem" of volunteer overload that is currently being experienced at the center. I replied today that I plan to attend every week, which I hope will not be too bothersome. It is probable, however, that part of my weekly shifts and likely entire shifts come midsummer will be spent in either the waterfowl or avian nurseries. During my first shift this week, after stating that I had discussed my intention to spend time in the waterfowl nursery this summer, I was reminded in no uncertain terms that this must be cleared with someone directly involved with the nurseries.

As the first part of the squirrel season is winding down, I am looking forward to my upcoming interviews with Phil Jenni, the director of the WRC, and Peggy Naylor, the main rehabber in Duluth and founder of Wildwoods Rehabilitation. Although Minnesota is blessed to have such an outstanding facility and staff as the WRC, the statewide distribution (as in most states) is weak. For example, St Louis County, the largest county by area in the state and one of the largest in the nation, does not have a single rehab-specific facility. Rehab efforts are increasing and improving in Duluth, and a fundraising initiative has been made for the facility. Limited capacity currently exists for rehab services, however, as all rehab in St Louis County is now performed in basements and backyards. The limitations of such an arrangement necessitate constant requests for transport of patients from Duluth or farther away to the WRC, with accompanying drives of 241 kilometers or more necessary. My understanding is that there is at least reasonably good communication between the rehabbers in Roseville and Duluth, and ideas for a statewide network of rehab centers, with the WRC as the "hub," have begun to be discussed. The prospect of such a system for not only Minnesota's wildlife but potentially that of other areas in the United States and beyond is exciting; but the extent to which such a hub-and-spoke-type model can and should be generalized is up for debate. While it is likely that the basic structure of such a model will, at least in principle, be broadly applicable, the treatments and services provided for wildlife will vary according to the species common to a particular region, as well as the financial and resource richness of the region.

7.

Given the rather severe shortage of squirrels currently in the nursery, I forewent volunteering Tuesday of this week and agreed to be put on a rotational schedule of every other week until we come out of the mid-season slump. I am currently trying to construct a schedule (with the volunteer coordinator—heeding last week's directive) for fitting in work in the waterfowl nursery this season. My hope is that this will not be a problem, since I have been told that the bounds of the waterfowl breeding season tend to fit roughly within the upper and lower limits of the first and second annual squirrel breeding seasons.

An observation of note that I had previously neglected to mention is my discovery of a dead gray catbird (*Dumetella carolinensis*) last week outside of the Forest Service Northern Research Station, where my advisor, David Bengston, works, and where my office is located. I surmise that it was killed from striking the window while in flight. While many other buildings on the St Paul Campus of the University of Minnesota have reflectors placed on them specifically designed to discourage such collisions, I have not noticed any on the Forest Service building, which was recently re-sided. I will try to remember to bring this problem and potential solution up to someone who could authorize the necessary changes. It would be great to institute a policy for all Forest Service buildings—perhaps even all buildings of the federal government—to have reflectors or similar preventive measures installed.

According to a whiteboard in the hall outside the squirrel room, the WRC has now released its 200th squirrel this season! This is exciting, especially since the season is only half finished. A question I have found myself struggling with in the past is raised by this number, however: Is it an indicator of a successful beginning to the season, or of tragedy? Something to be celebrated, or regarded with resignation (or even mourning)? After all, the success rate for rehabilitation centers hovers around 50% (although I don't have specific data on the WRC). This percentage, moreover, does not consider post-release survival. So, were the 200 released the fortunate members of roughly 400 admits, half of whom did not survive? I do not know, but I am interested in finding the answer. What is more, how many of these 200 will survive more than a month in the wild? 100? Given the current dearth of post-release survival studies, this question cannot be answered accurately. Perhaps survival rates are not of paramount importance with members of plentiful species; perhaps the more important reality is that an attempt was made, and an opportunity given. These questions, however, seem to point to the larger question of how increases in numbers of treated individuals ought to be interpreted. While common sense dictates that the number of animals in need who currently find their way to a rehab center are a proverbial drop in the bucket of the total existing number, and therefore that increases in total numbers of admits can only be cause for celebration, it is conceivable that a "saturation" point may come in the future after which an increase in number of admits is an indicator that more animals are being injured, orphaned, etc. But presently, are increases in numbers of admits more indicative of greater harm or increased compassion? This is certainly a question for future study.

I had the privilege of releasing two red squirrels today, along with a 13-lined ground squirrel (*Ictidomys tridecemlineatus*)—the latter species being the inspiration for the University of Minnesota's inaccurately named mascot, Goldy Gopher. These were the first of each species I have released, and I was quite excited to do so. The reds first had to be caught from a tall, narrow, rolling cage, and transferred into a pet carrier (no wooden box this time); this went surprisingly smoothly. My recent observations that the ratio of reds to grays in the nursery seems to be increasing and that reds nearing release-readiness tend to be more tolerant of human handling than equivalent grays appear supported thus far. After transferring the squirrels, we cleaned and restocked the rolling cage with food and water, then transferred all four reds from a self-feeding cage to the rolling cage. Although somewhat difficult to catch, these four were nice enough to handle with cloth gloves—except for the final one, who delivered a very minor bite.

I tossed a few nuts in the carrier to tide the soon-to-be-released reds over on the short journey. The "gopher," like last week's rabbit, came to me prepackaged in a taped-up cardboard box. In a manner reminiscent of the conveyed importance of oak trees to grays, I was told that it was critical to release the reds in a habitat with stands of coniferous trees. I immediately had a good place in mind and released them in the Reservoir Woods in Roseville in a stand of red pines (*Pinus resinosa*). Miriam and I only had a short walk from the parking lot, up a dirt trail and across the main paved walking trail. Interestingly, one was bright red and the other almost gray—so much so that it almost looked like a red-gray hybrid. I have not heard of such hybridization occurring with squirrels, but I would not be surprised if it does occur on occasion. Upon opening the metal lattice door, they showed reluctance to exit for a few minutes, then slowly emerged and scampered off in opposite directions.

After this apparently successful release, I was excited to see and release the ground squirrel. Although I had what I thought was a fairly accurate mental picture of what the species looks like, I was reminded that such archetypal images as can be found in field guides (or perhaps from the first or otherwise most impressing viewing of a species member, whether live or stuffed) are merely idealistic representations, and that concrete examples of intraspecies variation can often strongly challenge stereotypes. After a somewhat longer drive than expected—the directions said St Paul, although I believe it was Mendota Heights—we released Goldy in a large, hilly front yard. Although the box it was in read that the reason for its admittance was a dog attack, and I could see a doghouse adjacent to the home, I hoped it had learned to stay away from dogs. It was smaller than I expected, long and skinny, a bit like a cross between a chipmunk (genus *Tamias*) and a weasel (genus *Mustela*). I was surprised at how much it looked like a chubby snake as it lay low in the grass, and Miriam asked whether this might be a defense mechanism. Satisfied that we had done our job, we left Goldy with well wishes and another chance.

8.

I opened the door to the nursery today greeted by a loud squawking, reminiscent of the avian nursery. Being a bit tired and perplexed, I wondered what sort of strange squirrel could make such an abrasive sound. As I approached the area I perceived the sound to be coming from, my immediate impression of its avian origin was confirmed; there were several baskets containing birds in the back corner of the room. For a moment, I wondered whether I had entered the wrong room, but a closer look at the other baskets revealed their squirrelish contents. It appears that what just last week was a mammal nursery was now a mixed mammal/bird nursery. While my initial assumption was that these were simply overflow from the avian nursery, upon closer examination I noticed that "pox" was written on many of the identification cards. Soon after this a young woman from the avian nursery, a crew leader I recognized from last year, came in, apparently to check on the birds. I asked if they were all pox birds, and she replied that all five were (but that some with mites—which are very contagious—had been previously placed in this "quarantine" room, as well). Apparently, pox is not contagious, or at least not highly contagious, to squirrels, and hence their removal from the avian nursery.

We have begun the rotational shifts in the squirrel nursery, meaning for Tuesday there are three volunteers, and one crew leader scheduled for each shift until it picks up again. Although only two volunteers showed up today, there was very little to do. There are now fewer than a dozen baskets and cages occupied by squirrels. Although the formula is normally mixed by the time I arrive, I was apparently the first one at the nursery this morning. When the crew leader arrived, she told me the correct ratio for the mixture (one part powder: two parts water) and the importance of using the "weaning" formula for the open-eyed squirrels, as well as refrigerating and labeling any extra with the date and time of mixing. Although I did not expect there to be any grays in the nursery today, there were a couple—one of whom was a very fussy eater, protesting the syringe with growls and closed mouth. Expecting (and, given the small number of patients, being okay with) a long feeding session, I sat down with the squirrel and fed it its six milliliters of 75%-strength formula in tiny increments.

The shift leader voiced several very interesting observations today, which I will try to recall. First, she remarked that she was unsure as to where the gray with the spinal/head trauma was that had recently been in the nursery. I remembered this squirrel, and thought it must have died, been euthanized, or had a very rapid recovery. Although the latter might seem unlikely for most animals, it might be less unlikely for gray squirrels (bearing in mind the seemingly miraculous recovery of the gray with paralyzed hind limbs I had brought in previously, as well as a remark I had heard about temporary paralysis not being uncommon in grays). Second, she similarly questioned the whereabouts of a flying squirrel that I had not seen, but was sad I had missed, given I have never seen one—probably partially due to their nocturnality. Third, the shift leader remarked on a recent incident where a fox squirrel, the largest of North American tree squirrel species, had been released much too early after admittance. She stated that it was merely the size of an adult gray squirrel and had only been in the outdoor cage for three days—much too short a time for such a youngster to learn adequate foraging behavior. This was brought up to the mammal nursery coordinator, and apparently there was some back and forth between the nursery and veterinary staff concerning who was responsible for making the release decision. In addition to this anecdote, the nursery coordinator asked the shift leader and me if we would be willing to release some squirrels today. I replied that I would except that I had ridden my bicycle, and the crew leader said that she was no longer comfortable with releasing squirrels due to the uncertainty of where and when to release them, that it "stressed her out too much." Fourth, the shift leader alerted my attention to something I had previously been unaware of: the "orphan outreach program." I was told that this is a program whereby people with large amounts of free time can care for very young, orphaned squirrels, and that after they have reached a certain weight they are brought to the squirrel nursery. This helps explain why I had seen so few very young squirrels with eyes still closed in the nursery. She was unsure whether a rehab permit is necessary for such work. Fifth, my attention was drawn to something I am a bit surprised I had not noticed previously: The tables and counter tops in the squirrel nursery are heated. In a different manner from the avian nursery, where many of the baskets are set directly on heating pads, very thin plastic electric heating sheets (much like the devices used to defrost the rear windshields of cars, the coordinator told me) are placed on the tabletops, then covered with linoleum and/or contact paper.

9.

Two new "things" this week: Shift leader, and Virginia opossums (*Didelphis virginiana*) (in addition to the semi-new pox birds). It is beginning to seem as though expecting the unexpected is warranted for the so-called squirrel nursery. The new crew leader said that she is technically a co-crew leader with the expected person but had been unable to work in the nursery until now because of summer classes (she is a pre-veterinary student in the University of Minnesota's Animal Science program). Upon seeing the roughly doubled number of baskets from Tuesday, I at first thought that the second part of the squirrel season was somehow already underway but was told that they were opossums. I was also told that I'm "technically not supposed to look at them"—presumably because, as prey animals, it frightens them—but I had a hard time not looking at North America's only extant native marsupial. As is common with wild animals, opossums become more aggressive with age; this is thought to be an important trait for survival success. Many domesticated species, such as dogs, seem to exhibit a more constant level of aggressiveness throughout their lifespans.

It is widely held that those species that are domesticable possess the capacity for prolonged neoteny, or expression of juvenile traits such as docility, in their gene pool, and that this is an important explanatory factor for their evolution. Both neoteny and wildness can apparently be double-edged swords, however. On the one hand, animals that are more docile and trusting might have a better chance at getting along with humans and perhaps currying their favor. On the other hand, however, this lack of wildness makes them more prone, at an individual level, to use and abuse by insensitive humans. So, while the capacity for being domesticable may be seen as a species or "numbers" benefit (judging by the number of cows [*Bos primigenius*], chickens [*Gallus gallus domesticus*], and dogs in the world), it also may be disadvantageous at the individual level, since any one tame individual is less likely to defend itself appropriately from potential harm. I will stop here on this topic, however, as the wild–tame continuum is a massive and massively complex issue that is beyond the scope of my knowledge and this work.

But neither can I seemingly get away from discussing aspects of it, either. For example, my hypothesis about the reduced aggression of red squirrels seems to have been weakened today, as I was bitten hard by self-feeding reds today—hard enough, for the first time, to make me bleed, despite the admittedly thin fabric gloves I was wearing. Interestingly, I heard little of the warning growls that characterize aggressive grays but felt every bit of the biting force.

I released my first eastern chipmunk (*Tamias striatus*) today in St Pau—unsurprisingly at this point, outside of a residence. Unfortunately, it was raining hard, but hopefully warm enough that the "chippie" did not get too chilled. In any case, it nearly immediately scurried out of the enclosed PVC pipe tunnel—already nervous about releasing in someone's yard, I hoped nobody would see the PVC and think I was planting a bomb—and ran under a parked car for cover. The outside of the box it was packaged in indicated that it had been attacked by a cat. Attacks from pets are a real problem for small wildlife, and I am beginning to wonder whether releasing attacked animals at the site they were found is good practice.

10.

I only participated in one shift this week (today—if you can call an hour of work a shift), since the Tuesday morning crew is currently on a rotational schedule. I fed three very hungry squirrels today: one red, two grays. The room is completely devoid of self-feeders. There are now more other mammals than squirrels in the nursery—mostly opossums, but some white-footed mice (*Peromyscus leucopus*) as well. An additional new admit is a very messy American crow (*Corvus brachyrhynchos*), with a large sign on its cage reading "Do NOT talk to the crow!" This is presumably because of the danger of imprinting and the possibility of mimicry.

I interviewed Peggy Naylor, the lead rehabilitator in Duluth, last weekend. The talk went well, and she gave me a tour of the home-based facility that is currently Wildwoods rehabilitation, as well as lots of useful information. Peggy and her husband are completely devoted to the cause of wildlife welfare, and their commitment has not come without personal sacrifice. I sincerely hope a larger, more specialized facility will come to fruition soon in northeastern Minnesota. Thankfully, there are currently some very exciting plans being discussed in Duluth for an unprecedented natural living community consortium, possibly including a senior living facility, a wildlife rehabilitation facility (for Wildwoods), area nonprofits (such as Hawk Ridge bird observatory), and community and wildlife gardens. It is a shame, however, that the forthcoming state-of-the-art facility is not yet in place, as early this week Duluth was devastated by some of the worst storms and storm-caused damage in the history of the state. In such situations, the human tragedy and devastation is often focused on to the neglect of animal suffering and death. Unfortunately, the Lake Superior Zoo was visited by such horrors, with the drowning deaths of many of its captive animals.

On a smaller scale and somewhat more positive note: Yesterday, after returning home from a long day of transcribing, upon letting Tonky out I noticed that she was staring unrelentingly into a basement window well of our apartment building. There is a broken plastic sheet over it on a metal frame, which keeps neither water nor animals out of the well. In fact, I now recall having seen rodent skulls through our basement windows in some of the window wells of our building. They are almost certainly there because some poor mice, chipmunks, etc. fell in and could not climb back up the slick metal surface. This is something that could be easily prevented by covering the wells more adequately, and I will seek to remedy this.

As Tonky goes over to the well often to sniff and look down into it, I did not think much of her behavior at first—except that this time, she would not leave. I decided to take a look myself and was surprised to see a live baby rabbit in the well, about the size of my hand. I at first thought I might be able to scoop it out with a shovel, but then realized the well was too deep and the rabbit too mobile. It appeared that I would have to lower myself down into it, which is what I did, being careful not to scrape myself too hard on the broken plastic edges. I just fit. The rabbit cowered in the corner, and I picked it up with little struggle or protest, lifted it out, and set it on the ground. It remained still for a few seconds, then its ears perked up and, thankfully, it dashed away to hide in a brush pile. For a moment, I was worried that I might require some rescuing but managed to heave myself out without too much trouble. This was the first rescue I recall performing that did not end in rehabilitation, and it felt good.

11.

This morning, I had my first shift in the waterfowl nursery. As much as I dislike waking up before 8 on a Sunday morning (or any morning, for that matter), I was rather looking forward to the new experiences, as I had recently been told by an advanced intern in the mammal nursery that the waterfowl nursery was the most enjoyable of the three to work in. At this admittedly early stage, I am doubting whether I agree with her. This may be due to lack of communication between the crew leader and me more than anything else. At the end of the morning, I felt as though I had learned very little of the procedures involved with the opening shift. I am optimistic that next week will be better, however, as today's crew leader stated that he was unsure whether he would be there next week, indicating the future possibility of a leader with a more forward style of communication.

Another reason for the lack of immediate positive impression is the constant necessity of cleaning poop off *everything*. This includes brushing poop off towels (which nearly made me vomit), and scraping it out of food dishes, off one's rubber boots, and into neat little piles on the floor to be scooped into the garbage and washed down the drain. I thought for a moment that these activities would be prime candidates for the television show *Dirty Jobs*, then noted that for something to be considered a "job" in the relevant sense, one must get paid for doing it.

The layout of the waterfowl nursery is very different from that of either the avian or the mammal nurseries. The first and most noteworthy difference is that water is everywhere, compelling the wearing of rubber boots. A hose with a high-pressure spray nozzle is perhaps the main instrument of the room and is nearly always in use by someone for spraying the floor, pre-cleaning dishes, or filling water troughs. The young birds (not all are duckling— indeed, not all are even waterfowl) are kept elevated above the floor in long, open-topped wooden boxes covered by mesh fabric, and are equipped with heat lamps and suspended feather dusters, which the birds flock to if they feel threatened. Each one of these rectangular boxes is occupied by from one to about a dozen birds. By far most birds in the nursery are mallards, although there are also wood ducks (*Aix sponsa*), wild turkeys (*Meleagris gallopavo*), and pheasants. I was told that some other species of duck, such as common goldeneye (*Bucephala clangula*) and hooded merganser (*Lophodytes cucullatus*), are kept in a different room.

Fledgling ducks are kept in knee-high, plastic framed enclosures with camouflage netting on the sides. This netting is also kept over the outdoor cages, which contain plastic children's swimming pools filled with water. The indoor enclosures have large rectangular tubs of water in the middle, set on a green mat reminiscent of the surface of a miniature golf course; large circular water troughs are also included. Although I am not yet familiar with feeding procedures, as I spent most of the shift today washing dishes, I observed dry pellets, duckweed (genus *Lemna*), and crickets (genus *Gryllus*) being fed to the birds. It is my understanding that, unlike in the other nurseries, even very young ducks are self-feeding. The lack of need for handfeeding probably accounts in large part for the shorter maximum shift time— three hours, versus four in the other nurseries.

Large numbers of ducks—a dozen or more—are kept in each indoor enclosure. It was suggested today by an intern that such crowding, as well as dietary deficiencies, might contribute to a condition referred to as "angel wing," where ducks' wings are deformed. This caused me to wonder what a proximate cause of such deformity might be—perhaps elevated levels of stress hormones from overcrowding? The intern mentioned that it would be good to try to "space them out more," but I'm currently unsure as to if and/or how this will be accomplished. Indeed, stress seemed to be something of a theme of the shift. Upon arrival, the first thing I and the others noticed was an adult Canada goose (*Branta canadensis*) in the nursery. Nobody knew where it came from or how it got in, but it proceeded to hop over the low plastic "fence" of the fledgling enclosure and drink the ducks' water. This clearly stressed and frightened the ducks, who ran in a panicked clump to the opposite corner of the enclosure. One of the workers threw a towel over the aggressive goose and carried it out of the room.

Later in the shift, when most of the work was finished, the remaining four workers from our shift (including myself) walked over to the intake area to administer, and observe the administering, of meds to a duckling. The volunteer trainer walked over and said to the woman holding the bird that she should cover it up, since it is stressful for it to have so many people staring at it. The only response she got was an expression indicating disagreement, so she proceeded to state, "You look like you don't believe me…or that you don't think I have any business telling you this." The expression did not change, but she tentatively placed her hand over the duck's head. The trainer paused and said she'd go get something to cover the bird with, came back with a hand towel, and wrapped it up. The exchange made me feel uncomfortable but highlighted well the difficulty of keeping standards of practice uniformly high in volunteer-based activities bordering on professions.

12.

I have decided to collapse my formerly separate mammal and waterfowl nursery postings into a single weekly posting, unless or until significant events occur in the mammal nursery that necessitate a more thorough description. Currently, the activity in the mammal nursery is very low; last week there were three patients total in the nursery, rendering a 1:1 ratio of squirrels to caretakers. The upside to this is better care for each individual patient, but the downside is lack of activities for volunteers to engage in.

My time in the waterfowl nursery this week was more positive than the last. I took more of an initiative to ask questions, rather than simply observing and waiting for something relevant to occur. Even though I was about 10 minutes late to my shift, I was the first volunteer to arrive, and, like last week, the most obvious immediate task to accomplish was dishwashing. I do not mind washing dishes, but doing so means a good portion of my shift will consist almost solely of participation, devoid of meaningful (at least visual) observation, given that one must face the wall. I was very grateful, however, to have not been on towel duty, and the next time this is necessary I will wear a mask.

A different, more communicative intern was in the nursery this week, and I tried to take advantage of her communicativeness to ask questions. Upon arriving, I overheard her and another worker lamenting the lack of towels in the nursery, along with the fact that the dryer was broken. I sincerely hope the dryer gets fixed soon, as the amount of laundry that must be done on a daily basis is immense, and even with a functioning dryer basket overflowing with dirty laundry often line the hallway. I have faith that it will get repaired in short order, as I have been told that the volunteer trainer is the "go-to" person at the WRC for maintenance issues, and that has numerous connections for getting necessary goods and services with efficiency. Thankfully, we somehow managed to procure enough towels for the shift.

I visited the outdoor waterfowl cages for the first time today. The mallard cage is a large enclosure directly outside the nursery on the south side of the building, with plastic swimming pools instead of smaller tubs, filled with water and sprinkled with duckweed. I filled three large, circular plastic feeding dishes about halfway with feed pellets, and placed one in each section of the outdoor mallard cage. These are the "starter" pellets; perhaps they provide adequate nutrition for very young birds, but not juveniles (or at least not as most of their diet). Could this be part of the explanation for the mysterious duck angel wing epidemic? Interestingly, I was told today that sometimes ducks not in captivity, but fed by humans, develop this deformity of the primary feathers. If this is the case, then it seems likely that the causal explanation is more strongly diet- than stress-based. In any case, I was told that the success rate of wing-wrapped ducks with angel wing is very good, if a proper "figure eight" wrapping technique is used. I was also told about another common duck wing condition, "drooping wing," where a blood supply is provided to the feathers, thereby making them droop because of the extra weight.

A volunteer asked why all the outside ducks were females, and was told that both sexes were present, and that they look similar because they are juveniles. It was pointed out later that one might distinguish between juvenile males and females by the males' blue wing feathers. The older goldeneyes and mergansers are kept in cages near the mammal cages, on the east side of the building. I asked why the less common species are kept separate from the more common mallards and wood ducks, and was told that it was due to both space constraints and the stress more crowding would be likely to cause for wood ducks. Wood ducks are easily stressed—the rabbits of the waterfowl nursery, in this regard—and this is the primary reason for the posting of "interns only" signs on the wood duck area of the nursery. I do not recall if the duckling being administered meds last week was a wood duck or mallard, but if the former this helps explain why the volunteer trainer was so adamant about covering it up. Also, an intern brought in a juvenile Canada goose today, so apparently geese are intentionally kept at times in the waterfowl nursery.

13.

Tuesday of this week was different from normal. There was only a single squirrel, a gray, in the nursery, and since it was a "one-a-day" it was not to be fed during my morning shift. It initially seemed, then, that I had ridden over only to not feed anything. What is more, three other volunteers showed up, which meant severe overstaffing. Thankfully, the dryer was by then in working order, so we were able to make some use of our time by doing laundry. Once all the machines were full, we proceeded to the cage wash—as usual, there was no lack of work to be done there. After a round of washing cages, I was surprised and happy to be made aware by our crew leader that there was cleaning and feeding to be done in the flight room with birds normally being cared for by the animal care crew (ACC) (our crew leader is authorized to work with ACC patients). The ACC is charged with caring for injured animals year-round, and the flight room is, rather confusingly, divided between ACC birds and avian nursery charges.

For the first time, I prepared food in the general food prep area. This included slicing grapefruit for a Baltimore oriole (*Icterus galbula*), sifting mealworms, and mashing wet dog chow. The food prep area is at times a very crowded place, and I often had to wait my turn for some counter space. As an aside, I noticed the part of the enclosure designated for swallows had perhaps a dozen birds in it. Although "barn swallow" (*Hirundo rustica*) had initially been typed on all the identification tags, many of these had been crossed out and "cliff swallow" (genus *Petrochelidon*) written in their place. Despite this supposed correction, because of the shape of the birds' tail feathers I wondered if in fact they were either species, or perhaps bank swallows (*Riparia riparia*) or tree swallows (*Tachycineta bicolor*). It may make no practical difference, but I am curious enough that I might bring in a field guide for identifying purposes.

Because of the nearly complete absence of squirrels Tuesday, I did not make a trip over to the WRC on Thursday.

Sunday, like Tuesday, was a departure from the norm; only this time, there were too few volunteers in the waterfowl nursery. I, in fact, was the only volunteer, although there were two interns. In keeping with the trend, both interns were new to me (and, it turns out, one was new to the waterfowl nursery as well). Work seemed to progress very slowly, and by the conventional end of the shift at 11:00 AM, there was still plenty more to be done. To make things more difficult, the new intern left at 11, meaning that I was inadvertently placed in the position of being expected to stay late. I did not really mind, especially since I have left so many shifts this summer early because of lack of jobs to do, but I was beginning to get tired. Positively, however, I had the good fortune of a friendly and instructive crew leader who showed me how to perform many new tasks.

I was taught the procedures for cleaning and feeding in each of the three main types of enclosures for birds of different ages, which was a welcome change from doing only dishes and laundry-scraping (although I did each of these during this shift as well, being sure to wear a mask for the latter). Very young (up to a couple weeks old) and injured birds are kept in elevated, rectangular wooden boxes, which much resemble baby cribs with boards instead of bars on the sides. The food and water dishes are removed, and the insides of the "cribs" are sprayed down with a hose; care is taken to make sure the water is cold and the nozzle on a shower-like spray setting, so as not to injure the young birds if they get in the stream of water. A towel is folded in half and placed over half of the crib, and two food dishes, filled with dry starter food pellets and duckweed or shredded romaine lettuce (*Lactuca sativa*) in water, respectively, are placed, as well as drinking water.

Older, uninjured birds are kept in low-walled PVC pipe enclosures, the younger in this group indoors and the oldest outdoors adjacent to the building. In each case, all the furniture, mainly including pools, bricks, felt flooring, and rubber floor tiling, is sprayed down. I observed many instances of two floor tiles with textured surfaces tied together, then placed over the side of the pool in the shape of a house roof. I assume this is intended to help the ducks enter the pool, and I indeed saw many using these makeshift ramps. Others, however, seemed to have no trouble hop-stepping right over the side. Care is taken to not get an excessive amount of wastewater into adjacent enclosures when spraying and emptying pools, but some minor flooding is unavoidable. The same types of food are given to these older ducks, but the duckweed or lettuce is sprinkled directly into the pools.

Outdoors, ducks must often cope with higher temperatures and more direct sunlight than those indoors. It is critical that they have full pools for cooling, swimming, and bathing, but when I entered the outside cage, I discovered that two of the three pools only contained a few centimeters of water. Upon filling them I found that they both leaked, so we replaced one with a new pool that was on site and the other with a pool we found that was patched with duct tape (we put more tape on before filling it). A sprinkler is kept in the cage that can be used on very hot days. For the older indoor and outdoor ducks, I was encouraged to periodically spray them with the hose. They flap their wings in response, providing some exercise for the all-important wing muscles.

An incident of confusion, ending in unfortunate news, occurred Sunday. A woman in scrubs, who I believe to be a vet tech, entered the waterfowl nursery asking if I knew of any woodpeckers there. I said that I did not, and we both thought it strange that any would be in the waterfowl nursery given they are normally kept either in the avian nursery or in an outdoor cage. She left but came back later after realizing that she had misread some messy handwriting that referred to a wood duck. The crew leader then entered the nursery and, after being questioned as to its whereabouts, replied that she had found it earlier on its back and hypothermic. She had placed it in the oxygen room, thinking this might help it, but unfortunately it was found dead shortly thereafter.

14.

This has been a rehab-less week, as the nursery is now entirely devoid of squirrels. I received a phone call from one of the shift leaders on Tuesday; she told me that it would likely not be worth it to come in. I had not been planning to come in since I was not on the shift rotation list this week that had been sent out by the other shift leader. Unfortunately, this list had been emailed to the volunteers but not the co-shift leader, so upon request I sent it to her. This sort of poor communication between shift leaders does not seem altogether uncommon.

In addition, I skipped volunteering with the waterfowl this week because of a race I participated in (I gave the volunteer coordinator notice in advance). Despite this dearth of volunteering on my part, it has not been a rescue-less week. For the second time this summer already, a baby rabbit was discovered in our egress window—this time by Tonky and Miriam. Miriam reported that it looked like it had hit its head, but as I lowered myself into the well, I instead found a large, swollen tick, along with several smaller ticks. I had not yet contemplated or encountered any opinions on whether wildlife rescue should include a complimentary parasite removal service, but I pulled the largest of the ticks off. Worried that the others might be deer ticks (*Ixodes scapularis*) because of their tiny size and therefore potentially carriers of Lyme disease, and not wanting to stress the bunny too much, I left the rest and lifted it to the ground. It hopped away vigorously, into the neighbors' yard.

Deciding it was time to do something about this death trap, I rummaged around in the garage for something to put over the large opening but found nothing suitable. I called my landlord, and he was there the next day, putting new covers, alas, on all but the problem well (that he at least was covering the smaller wells was good, however, since I had seen many rodent skulls in them). I explained that the large one was the most problematic, and needed covering the most, but he replied that he could not put any permanent or heavy coverings on it, since the basement resident needed a means of escaping directly from his unit in case of fire. The best resolution we could come to is that we would think about a solution. Until one appears, I will try to monitor the well closely. This situation, unfortunately, demonstrates well a way in which technological arrangements that are beneficial to humans can be disastrous to wildlife. Animal advocates and conservationists must not ignore such issues, as they are already prevalent and will only become increasingly so in the future.

15.

The second "batch" of squirrels has yet to come in. The waterfowl nursery, however, is going strong in terms of patient numbers. A father-daughter duo, who I see about every other week, was volunteering in the nursery today. These are the first obviously related people I recall having seen in any of the nurseries. I find myself wondering about their motivations for volunteering as (presumably) being a middle-aged man and teenage girl, they each seem to be members of underrepresented demographic groups at the WRC. The man is a bit gruff and taciturn, while the girl seems confident and self-assured. The intern—yet again, someone I don't recognize—was apparently training in a new participant today and seemed content with leaving the majority of the indoor work to the more experienced volunteers.

As I was cleaning the first of the indoor divided enclosures, the girl, seemingly taking on the role of the intern in her absence, told me that I should not bother removing the food and water containers as I had just done, since they could just be refilled while in the enclosures. I had assumed that both would need to be washed, but then remembered that the large, heavy water dispensers are often left where they are and filled with a hose (the food dishes, on the other hand, certainly needed to be washed, as ducks often use them for toilets as well). I was a bit annoyed at being bossed by a teenager who seemed to have little respect for my seniority of age and degree, but remembered the special context I was in, where textbook knowledge must often bow to the dictates of direct experience and established procedure. I am in my rather inglorious position for certain reasons, I reminded myself, namely, to participate, observe, and care for needy creatures, but not to seek to bolster my pride by appealing to credentials. I wondered whether it was good practice to consistently not wash these communal water dispensers, then lifted the container back into the enclosure.

16.

For the first time this season, there was an intern in the waterfowl nursery that I recognized. This shift went quickly and uneventfully. I am noticing more empty enclosures. We have apparently been blessed with a bounty of duckweed, as I have not seen any lettuce substitute recently and the cat litter turned duckweed container seems full most of the time. Unfortunately, it was also frozen today, and I scraped skin off my knuckles trying to chip away at the frozen mass with a screwdriver. It would be interesting to know what proportion of objects at the WRC have been repurposed from intended human uses; my guess would be the majority. Although I am aware of a substantial literature on the unintended consequences of technology, I am unsure whether much has been written on unintended *uses* of technologies. The WRC might be a good place to start such a research program.

I constructed a makeshift window well "bridge" this week with some pieces of plastic siding and cardboard I found in the garage, hoping these materials would both be light enough to keep me out of trouble with the fire marshal, and strong enough to allow small animals to walk across without the structure collapsing and falling into the well. If the latter does occur, hopefully it will be enough of a visual disturbance to attract someone's attention in time to save the animal. I had never thought of myself previously as having engineering potential, but this latest project has me wondering.

I have noticed a bizarre-looking rabbit under our bird feeder recently, which had me rather concerned. It appears to have a horn-like projection growing from an ear. I recalled hearing the director of WRC, Phil Jenni, addressing concerns over lesions on rabbits in a recent public radio broadcast. Someone had asked what was wrong with them, and whether they should be killed. I do not recall exactly what Phil said in response, but he advised that the animal was likely not a threat and should not be killed. After doing some quick research, I discovered that the "horned" rabbit under my feeder likely has the cottontail rabbit papilloma virus (CRPV). Although it is not usually dangerous, in some cases it can interfere with the infected rabbit's ability to eat, possibly causing starvation. Rabbits infected with this virus may have been the historical inspiration for cryptids such as the jackalope and wolpertinger. One can only imagine how many past monsters have been conjured up out of encounters with infected but otherwise normal individuals.

17.

Today was my first shift in August this season, and the transition to a new month was marked by an accompanying transition to a nearly empty waterfowl nursery. The contrast to last week was a bit shocking, as only one of the three indoor duck enclosures was occupied, and only four brooder boxes (I was happy to finally learn the technical term for what I had previously thought of as bird cribs). Because of this dearth of ducks, coupled with two working interns, the work was relatively light. Unfortunately, I made more work for myself by accidentally cracking a "pool" (a rectangular plastic container likely not designed to hold water) from dropping a brick into it, rather than placing it in gently like I should have done. Luckily, I was able to find an unused, uncracked pool to fill, as well as some hot pink duct tape to patch the damaged one with.

A glance at a chart on the nursery door revealed the death count so far this season: two mallards, two wood ducks, and one wild turkey. Each was found dead in its cage/enclosure (DIC), euthanized, or died after being placed in the oxygen room. I have not yet directly witnessed any dead birds in the waterfowl nursery.

I made a trip to the supply room today, in hopes of finding some pelleted waterfowl feed. After first looking in the walk-in cooler, where I discovered many cardboard boxes of fruits and vegetables, as well as milk replacement mixes for foxes and smaller mammals, I found several stacked bags of the feed out in the main storage area. After pouring the entire bag into the appropriate plastic bin, I sent it up in the dumbwaiter, which stops at the end of the hall just outside the waterfowl nursery. Although operating the dumbwaiter seemed straightforward, I was greeted with a loud and enduring "beep!" upon first opening the door, after which I hastily closed it and sent the feed up, hoping that nobody would rush over and scold me for misusing the equipment.

A copy of *Wildlife Rehabilitation Bulletin* had been left on one of the tables in the break room, which I was happy to see, for the first time, in print. As I expected, many, but not all the articles were technical explications of veterinary procedures. In the book review section was a review of a book, *Wildlife Search and Rescue: A Guide for First Responders* (2012), which I am quite interested in reading.

I held a wood duck (aka "woody") duckling for the first time today as it was being re-banded. The strength and vigor of struggle this little bird displayed was unexpected, and I nearly dropped it. One of the interns told me that the ducks must be re-banded frequently for purposes of identification, as they grow quickly. If the band is on the leg too tight, it can cause circulation difficulty as well as discomfort. A rule of thumb for band replacement is if the band cannot rotate freely on the leg, a larger one is needed. In addition, I caught an adult mallard for the first time for the purpose of taping its drooping wing. After having seen how aggressive mallards can be, as well as the wounds their dull-looking beaks can inflict, I was a bit wary, but thankfully this one intended me no harm. I was told that wing tape is removed every three days. If the wing(s) looks normal, the duck is released into the larger outdoor area; if an abnormality is still detected, it is re-taped and kept in a smaller recovery area.

18.

Because of the near emptiness of the waterfowl nursery, I did not volunteer there this week. I did, however, work in a surprisingly full squirrel nursery. Many squirrels still had their eyes closed, and therefore required feeding with the "eyes closed" formula. Expecting all eyes closed squirrels to be tiny, I was surprised to find that many of them were larger and heavier than the smaller of the squirrels with eyes open. I am unsure as to what causes the eyes to open.

There were squirrels that needed feeding four, five and even six times per day, meaning that some of them had to be fed twice before the AM shift was over. This did not seem overly laborious to me, as it still paled in comparison with the frequency of feedings required for the avian nursery. What I am sure was overly laborious was an action I am not authorized to take—injecting dehydrated squirrels with LRS. It had been determined that several squirrels had unfortunately missed their last feeding through negligence of some sort, making it more likely that they would be weak and dehydrated. Added to this was the reality that virtually all of them had been very recently admitted, after likely having been devoid of parental care for an indeterminate period. Because of these factors, many patients were deemed dehydrated. A longer list than I had yet seen on the issue log was the result, meaning that the shift leader would be treating squirrels long after our shift had officially finished.

I am afraid that I have lost my touch a bit with the syringe, and made several squirrels aspirate because of this combined with their eagerness to feed. Although I have been hesitant to use a suction bulb on squirrels, after observing such usage more and being assured that it is not traumatic, I attempted it on a squirrel during my Thursday shift. I sensed that there was something wrong with this squirrel; as I was holding it, it seemed to be breathing laboriously, in a raspy manner. In addition, there appeared to be dried mucus under one of its nostrils. The procedure seemed to work, with a significant amount of liquid apparently being removed from the patient's lungs. I notified the shift leader of this, and she replied that the squirrel was being treated for pneumonia. While sad (for the squirrel), this gave me some confidence in my diagnostic ability.

19.

On August 16 I received an email from the WRC's Director of Recruitment and Volunteer Services that the waterfowl nursery will be closing for the season on August 24, and that "there will still be a few duck(s) left, but there will be plenty of interns around to care for them." Therefore, reluctant as I am to get up before 8 AM on a Sunday morning in any case, I am afraid I have seen the last of the waterfowl nursery this year - at least in a regular volunteering capacity. I must admit that, of the three nurseries I have experienced during my relatively short time at the WRC, this was not my favorite (I am hesitant to strongly state which was my *least* favorite). However, I certainly learned some important and interesting things about duck health issues and treatments, as well as developed a significant research question: What is the cause(s) of the mysterious wing ailments, angel wing and drooping wing? Of all the indicators of potential gaps in overall rehab quality, this stands out in my mind as the most significant.

I did not volunteer at the squirrel nursery on Tuesday, since I was not on the regular shift list for this date. I also was delinquent on Thursday; I justified this by my feelings of illness that morning, as well as my memory of the crowdedness of last Thursday's shift with eight human bodies in the nursery. Recalling that Thursday's shift leader had requested help on the Friday morning shift, I made it to this one. Upon entering the squirrel nursery Friday morning, however, I was very sorry that I had not showed up Tuesday and Thursday. For some reason I had assumed that the second of the squirrel litters would be the less numerous, and that the modest numbers I had seen last week would not increase by much if any, but one glance told me that this was a faulty assumption indeed.

A basket of squirrels was on virtually every available above-floor surface. One of yesterday's volunteers, after seeing the numbers, had come in this morning as well. She told me that they were so swamped Thursday that they didn't even have time to wash dishes. I now felt very guilty indeed. As an attempt to somehow make up for my absence—or perhaps just to address the need that was then before my eyes—I stayed later than normal, although I had a long drive (and accompanying hike) out to a campsite planned for that day. The nursery seemed terribly disorganized and hectic, with no apparent rhyme or reason for the normally alphabetical placement of the baskets. An intern did not arrive until about halfway through the shift, at which point she stated that she thought one of the tables was for pneumonic squirrels, which would partially explain the disorganization.

Many of the squirrels I found to be much too large for the basket and feeding regime they were listed for, and I took the liberty to switch them myself as I had normally done in the past. It turns out, however, that this intern, who I had not worked with previously and who was apparently filling in for another intern, thought I should merely make suggestions about moving the squirrels up to fewer feedings or self-feeding. I surmised that this was because many of them had not even been in the nursery long enough to be on undiluted formula (the protocol is to move them up from either LRS or 25% formula concentration in 25% increments each day), even though they were well over the weight threshold for self-feeding. While this was a bit frustrating to me, I was glad that the interns were engaged in thoughtful decision-making and record checking. In addition, I was shown the specific procedure for requesting a vet check after having observed for the first time a squirrel that appeared to have blood around its anus. This process required a short trip to the intake area, where the patient's species, identifying marking, case number, enclosure number, and problem were recorded on a special form on a clipboard.

20.

Apparently, a call for help in the squirrel nursery has been sent out and heeded, because I counted an all-time high of 13 humans helping on Tuesday. It got to the point where the shift leader had to suggest that people leave, as it became nearly impossible to move about the room. Despite this surplus of help, we did not end much earlier than scheduled. Many of the squirrels I had identified as candidates for self-feeding last week had since been transferred to cages, and attending to these often ornery and aggressive patients took most of the time that had been made up on the baskets.

A spate of respiratory tract infections (RTIs) seem to have a death grip on the squirrel nursery right now, and I have noticed a significantly greater number of DIC (died in cage) and E (euthanized) designations than I recall for the first litter of the season. Roughly every other squirrel I pick up appears to show the symptoms of an RTI, by way of wheezing, sneezing, and/or mucus around the nostrils. Many are on medication or probiotics. As much as the squirrels obviously dislike such treatment, I try to draw out any liquid from the lungs through the nostrils with a bulb syringe. Sometimes, the amount drawn is astounding. This causes me to wonder about the extent to which aspiration of formula is a contributing factor. Many squirrels are strong enough that they can suck the syringe dry on their own, and any pressure on the plunger whatsoever will cause them to aspirate. Although it would be time consuming, would a policy of using the bulb syringe on every squirrel after every feeding improve the success rate? My hunch is that it would, but the question of practicability is important, as well as the possibility that more sharing of bulb syringes would enhance the infection rate.

There were fewer volunteers Thursday, but still a good number. Because of the high prevalence of infection currently in the nursery, an immune booster is now being added to the formula. This makes the previously white liquid speckled orange, conjuring up images of a squirrel Halloween party. I sincerely hope that this additive has the intended effect and doesn't merely dye the syringes orange as it looks to be doing.

An albino squirrel is currently in the indoor cages. It is remarkably nice for a self-feeder and allowed me to weigh it barehanded with no protestation. This squirrel is a true albino and hence a rarity judging by its pinkish eyes, as most "albino" squirrels are merely affected by leucism, a fancy term for whiteness. In either case, however, these animals are much more sensitive to predation, as at non-snowy times they are much more visible. There is even a student-run international Albino Squirrel Preservation Society, with chapters on several university campuses, to raise awareness of and advocate for these special creatures. One of the volunteers commented that a white squirrel used to frequent her property, and that she fed it peanut butter and molasses sandwiches.

I participated in what I thought to be a very strange event (or perhaps nonevent) while driving last week. Pulling into the alley behind my apartment, I noticed a goldfinch up ahead bathing in some standing water in the middle of the alleyway. I slowed and crept closer in my car, but it did not fly or move away, or even perceptibly change its behavior. Eventually, I got close enough that I could no longer see it over the front of the car. Not wanting to run over it, and curious to see whether it had moved, I backed up. It was still there. It did not appear injured, so I simply drove around it. Checking in my rearview mirror, it remained in its spot, splashing water over itself. Did it know that, had I continued my trajectory, I would have passed over it, probably clearing it harmlessly? If so, this was an astonishingly perceptive, energy-saving judgment for "just" a little bird.

21.

Given the relative lack of outstanding experiences this week, coupled with the nearing of the end of my documentation period, this seems like an appropriate time to reflect on some of the things I have learned thus far. Thinking back on the beginnings of my interest in wildlife rehabilitation, where has this path of inquiry and discovery taken me? This is a broad and difficult question to answer thoroughly. So much of what I have done has been lost to memory already, or never even brought to the level of consciousness to begin with. Is it possible to become the phenomenon of interest, be completely absorbed in what one is doing and what is going on, and remain conscious of the relevant details? This was supposed to have been an exercise in participant observation, but sometimes in the milieu of trying to save lives I am afraid that criticality has fallen by the wayside.

The special character of the process of becoming a rehabbing insider is that I am ethically obligated to perform the work I have been doing well, in a sense that is not present in many other studies using this methodology. Ironically, a large portion of participant observational and ethnographic studies have been on deviant and even criminal phenomena, such that performing the relevant work well would likely be seen as wrong, abhorrent, or vicious from the perspective of many outsiders. For example, one would hardly be morally reprimanded for failing to perform one's role as participant in drug dealing culture poorly (although one might be professionally or academically reprimanded). However, if I am distracted during a feeding session by clever and interesting ideas perhaps worthy of academic acknowledgment, lives may be harmed or even lost. So, in the setting I have chosen, awareness and observation of the whole setting must sometimes be sacrificed for the immediate and higher good provided by life-preserving tedium.

As for the writing process itself, it is truly turning into memory work. Part of this is my own fault, for waiting too long after sessions to write. At the end of this volunteer experience, I will have (barring catastrophic technological failure) attained my tacit goal of adding data to the exceptionally thin database on the human aspects of wildlife rehab. In hindsight, however, I see many ways I could have done it better. I suppose this means I have learned something after all, which is good but somewhat unsatisfying. Although learning is crucial for the future and primarily valuable because of this, there is always the temptation to want to say that nothing of value has been attained short of rewinding the "tape" and rerecording.

At this point, the days and experiences are beginning to run into each other, and what I once perceived as a neatly chronologically ordered log of events and experiences is beginning to feel a bit long and muddled. I started this journal with concise descriptions and reflections, but now it seems I am writing much more although, due to repetition, I feel I have less to write about. While not quite déjà vu, I am sensing the repetition of a cycle that I do not recall having sensed last season. This is undoubtedly at least partially due to the two squirrel litters in a single rehab season. I think it is also because of my enhanced familiarity with and habituation to the research setting. The more commonplace something feels, the less aspects of it seem worthy of notice or even noticeable. I find myself constantly on guard against repeating myself, although I know it is inevitable. In such a routinized setting as the WRC, the same types of events occur repeatedly: Admittance, feeding, medication, death, release, etc. It is often the minute differences between how individual instances of these events play out, however, that makes the very great difference between success and failure.

22.

This week has been crazy. Given the start of classes at the U of MN Tuesday, volunteer and intern support has been lacking. Worker numbers have been cut in half or more, with the number of squirrels remaining relatively constant for the time being. I was just told that we are "ahead of schedule" for the number of squirrel patients this year, which, despite my relative lack of experience in the mammal nursery, does not surprise me. Baskets are piled on baskets, and every basket-worthy space in the room, except for one in the back corner where squirrels whose cages are being cleaned can be temporarily placed, is occupied. For every shift I have been at during the last two weeks, squirrels still needed feeding at the stroke of noon. How might this late-season dearth of help be avoided in the future? Is there a way of anticipating and countering the mass exodus of student workers? Perhaps putting more effort into recruiting retirees or those without new commitments in the fall?

A question that has been on my mind, and that has come up in discussion during shifts, is the cause(s) of and health status of the second yearly litter, in general. Is its production merely an evolutionary tactic to buffer the potential catastrophe of a failed first attempt at breeding for the year? If this is the case, why is the "strategy" not more common among other species and taxa? Or perhaps it has nothing to do with selection or evolution and is simply a random occurrence. Aside from the cause of the second litter itself being born, why does it seem to consistently fare worse than the first, as was attested to by an intern? Is it, as she suggested, that the mother is comparatively nutritionally impoverished prior to giving birth to the second as compared to the first litter, and the resultant weakness and susceptibility to infection that often characterizes the later kits is caused by this lack of nutrients?

Tuesday was, as I alluded to above, hectic. People seemed stressed out, aside from the load of squirrels. Commands were being barked, orders shouted, seemingly to little avail. The squirrels almost appeared as if they sensed this stress, with one biting onto the shift leader's finger and refusing to let go. It was biting her finger lengthwise, like a tiny corncob. She hurried over to the sink and doused its head with water, at which point it relinquished its bite. Two redeeming experiences from Tuesday were my discovery of another true albino squirrel (younger than the one I recently found) and finding that a squirrel that had apparently fallen from a table the previous week because of my neglecting to keep the top on its basket while feeding a basket mate, appeared healthy.

It is both interesting and frustrating for me to acknowledge how the development of bad habits can override logical reasons for doing or not doing something. For example, it is easier to simply take the front clip off baskets instead of both clips and the entire top half, with the added benefit of decreased probability of escape. However, I had somehow—perhaps because this is necessary when the basket is to be cleaned, which is only relevant for "two-a-days"—gotten into the habit of taking the entire top half off every basket, whether it needs cleaning or not. I am happy to report that I seem to have broken myself of this habit now but upset that it took me so long to realize that it was a problem.

Thursday was perhaps the most stressful and arduous shift I have yet had in the mammal nursery, maybe even any nursery. Rather than the usual four hours of intense labor, this time it was five-and-a-half. Although previous shifts could undoubtedly have gone this long, I had not before today felt an obligation to stay later in this nursery. Part of this feeling of obligation, I think, derived from an attendant feeling of being a de facto crew leader. Continuing in the trend of losing interns to the start of the fall semester, I arrived today to find no official crew leadership. We (all volunteers) scrambled to figure out how to make formulas, which none of us recalled being taught. I was asked many questions today by other volunteers, providing me with an uneasy feeling of unwarranted authority. Someone noticed a dead self-feeder, and I reluctantly removed it and threw it in the intake room trash.

I received an email today requesting that I take over co-crew leadership of the mammal nursery on Thursday, given the normal shift leader's new school-related responsibilities. This was quite unexpected but flattering, and in an odd way seemed to validate my feelings of responsibility, however unmerited. It also helped me feel that, through all the shortcomings and perceived failures, I must have done *something* right after all. I somewhat reluctantly agreed to the attendant duties for the remainder of the season, knowing that this will significantly change my role and the study I am conducting. It will, however, provide a new perspective, which will be welcomed. If I did not perceive such a need for someone to take this position for the rest of the season, along with the possibility of developing a new insider viewpoint, I would be much more reluctant to agree to it.

23.

As I've alluded to many times previously, the pressure to accommodate and serve is being felt in earnest right now. Although numbers in the mammal nursery are on the verge of seasonal decline, crews are currently being stretched far past the limit that their current level of help imposes. Even so, the WRC seems to have a strong ability to attract volunteer support (although I am unsure at this point how well such support is retained); do other centers, and will other centers in the future, have similar levels of success? If not, what might be done to increase support?

A question that just entered my mind, after listening to a Radiolab podcast, is the extent to which rehab facilities are like and different from zoos. I am not prepared to attempt to answer this question now, but I think there is a promising possibility of rehabbing becoming a more ethical alternative to zoo-going and zoo-keeping. While the potential audience may be smaller (depending, of course, on the overall prevalence of rehabbing), the ultimate payoff in terms of freedom for animals is much greater, with the benefit of participants becoming interested in native wildlife, rather than the nearly exclusively exotic species that decorate many zoological gardens. Which brings up another, perhaps unanswerable question: How many exotic or invasive animals have originated from or been inspired by zoos? I expect that the number is staggering. What if our postmodern problem of animal invasion derives as much from zoos as from some sap in Central Park releasing house sparrows and starlings? The question of whether rehab centers can be all out replacements for zoos is complicated by the potential role of rehabbing for children coupled with the vehement opposition of many in the wildlife rehabilitation community to the use of animals for educational purposes. Should we be content with having dead animals on display in museums rather than forcing them to live out their lives in unsuitable environments, however hard staff might try to make them suitable? There is, however, always the option of using animals that cannot be released back into the wild for education instead of euthanizing them.

Interestingly, my attention was drawn by a fellow volunteer to perhaps the first "zoo-style" exotic I had yet witnessed at the WRC. I mean this in the sense that it was not an invasive species that has established a self-sustaining group, such as a house sparrow or nutria (*Myocastor coypus*), but an individual curiosity one would be likely to see on display. It was listed as an Aldabra tortoise (*Aldabrachelys gigantea*), supposedly escaped from a domestic setting. Leaving questions about how a tortoise escapes aside, what is the probability that such a specimen so thoroughly ill adapted to a temperate climate find its way to Minnesota apart from intentions of a person or persons to acquire and possess a member of an exotic species, whether as a pet or as an exhibit? This must not be a case, such as that of starlings, of release with intent to survive and reproduce in the wild. And what is a zoo animal but a pet of the public? The concept of a public pet brings up a host of ethical issues, not the least being who is ultimately responsible for the animals' care. There are many cases in which this responsibility seems to have been lacking, such as the recent case of the flood in Duluth where a strong contributing factor in the deaths of many animals was likely neglect.

If zoos are in the business of providing public pets (with, of course, the possibility for temporary holding and breeding of critically endangered species), what are we to say of what rehab centers provide? Perhaps the primary difference is the intentional provision of services to the wildlife themselves, in the latter case, instead of to humans. Although education, entertainment, and therapy undoubtedly accrue to those involved in rehabbing to varying degrees, these effects are presumably not the main reasons that rehab centers exist, or that rehabbers engage in rehabbing.

Today was a trying day, the first day as a (co-)crew leader. I was considerably more anxious than normal before the shift, and my nervousness was only compounded by the necessity of telling my fellow crew members that I was now one of their crew leaders. Did they take this as an indication that their work was somehow deficient compared with mine? Perhaps they felt as though they deserved the position more than I did, lacking in tangible benefits though it may be? I am not sure that drawing crew leaders from the volunteer pool late in the season reflects good policy, because of the possibility of hurt feelings and resentment.

The head of the nurseries came in to train us in around mid-shift. Although I was not aware of this at the outset, she later told us that she had been working at the WRC for around 14 years. We were shown where the medications were located, what types there were, how to interpret the potentially confusing med sheets, how to draw the meds up, and the importance of inserting the syringe far back into the animals' mouths (as many meds do not taste good). I found the med sheets to be fairly mind boggling at first, with boxes and fields for date, time, type of med, amount, etc., and many sheets with three or more types of med on them, sometimes progressing onto the back of the page. A confusing aspect that was pointed out is that there is often inconsistency between whether the drug name or brand name is provided, and drug descriptions often cannot be readily found. Also, the name provided is often merely an acronym, with the potential for confusion or misidentification. The usually liquid meds must be given in tiny amounts – hundredths of milliliters – and drawing them up into the syringe in the precise amount can be frustrating and eye-straining.

By far the least enjoyable duty that I am being compelled to perform is injecting the squirrels with fluids. I was hoping that I could defer to my more experienced co-crew leader in this area and avoid having to inject, but this hope did not come to fruition. Although I have not yet stuck myself with a needle, I have been told it is easy to do and have already witnessed a Tuesday morning crew leader stick herself many times. Before putting them on a syringe, the needles must be twisted and pulled out of a plastic cap, which is difficult and provides ample opportunity for sticking oneself. Then they must be fitted on the syringe, poked into a bag of LRS (which must be held upright), and the plunger must be drawn back with force and the syringe allowed to fill. Although air bubbles are not a severe danger since we are not giving intravenous injections, they should be flicked or pushed out if they are large. The dehydrated squirrel then must be held down forcefully, the skin on its back pinched up, and the needle inserted into its shrieking and straining body. The skin must be inflated with liquid until it becomes semi-hard. The goal is to get 5% of its body weight of LRS in—the same amount of formula it is fed at every feeding—but it seems that fluid begins leaking out the insertion hole before this point is reached, at which time the needle should be pulled out. This process must be repeated for every squirrel on the list, which can amount to dozens. I am glad this season is almost over.

A very interesting coincidence came up last week that I've just now recalled. In discussion of how we came to learn of the WRC with the person who, ironically, would become my co-crew leader this week, some striking similarities were found. Each of us first made our way to the WRC after finding an injured eastern cottontail being attacked by crows. Additionally, and unfortunately, neither of the rabbits survived. Although this similarity of experiences may be merely coincidental, it does not seem unlikely given that the yearly number of cottontail admits is nearly twice that of the next most numerous species (eastern gray squirrels). Thinking about the reason for this superabundance of rabbits compared with what seems to me an already staggering number of squirrels, I wonder how much of it is because of the inability of rabbits to climb, and the fact that squirrel nests are normally in trees whereas rabbit nests are on the ground. This makes rabbit nests especially susceptible to being mowed over.

24.

I became rather violently ill last weekend with flu-like symptoms, and so unfortunately, I missed my Tuesday shift. I wonder whether this illness is linked to the increased anxiety and stress levels experienced performing my new duties as crew leader. Alternatively, but relatedly, perhaps I contracted something from the squirrels or humans in the nursery. Squirrels are not known as being particularly zoonotic creatures; but I do recall the volunteer coordinator at the WRC having recently been out sick for what seemed like a relatively long period of time, and another volunteer stating that she thought she had West Nile virus. Speculation is probably worth little in this case, and it was likely a fall flu bug unrelated to wildlife that caused my ailment. It is worth noting, however, that there is now a box at the top of the WRC home page stating that incidences of West Nile in birds are currently on the upswing in the area.

Thankfully, I was able to make it to my shift as co-crew leader this morning. As I had hoped, the squirrel numbers have finally begun to recede. Now most squirrels left in the morning crew's care are two-a-days or self-feeders. Strangely, there was still a six-a-day, however, that was being kept in the ICU's incubator this morning. The first thing I did upon arrival was check on it, but unfortunately it was dead when I arrived.

Today was the first day my co-leader and I drew up meds and administered meds and fluids without any supervision or help. Things seemed to go smoothly in general, I'm sure at least partially because she is a great person to work with. It did feel a bit stranger, however, beginning to construct a routine from scratch with another worker. Successfully doing this often means tolerating and accommodating a certain amount of personal quirks and habits, which can be challenging; but thus far cooperation has seemed quite easy. While there was no shortage of meds to be drawn and administered, I was glad that there was not an overwhelming amount of fluids to be injected. This, unsurprisingly, seems very stressful for the squirrels, and they appear to greatly dislike the pressure undoubtedly caused by the hard packet of fluid formed in their backs during injection. They look strange and uncomfortable after the injection, as though they have little backpacks under their skin. Also, I am afraid that a needle might inadvertently puncture deeper than just the skin, given the vigorous protesting that often accompanies an injection. For these reasons, along with my perception that they were not sufficiently dehydrated to warrant the trauma of injection, I declined to inject two squirrels today who volunteers had marked down as needing fluids.

A select cage of self-feeders was very difficult today. Despite my belief (or perhaps wishful thinking) that this was not possible, an extremely aggressive gray bit through a leather glove I was wearing and gave me a small, bleeding puncture wound. This is not the first time I have been made bleed at the WRC, but it was the first time I thought the injury required some first aid apart from soap, water, and hand sanitizer, so I went to the treatment room looking for a bandage. The veterinarian, who seemed a bit annoyed that I initially could not find the first aid kit, pointed me in the right direction. She was anesthetizing a gray she had recently taken from the nursery, prior to an attempt to repair a split upper lip.

The biting incident made me very reluctant to continue with the self-feeders, but thankfully this cage was an outlier. The others I worked with were almost strangely docile, including a 250+-gram giant who was not the least concerned with my handling him. It appears that we are in the process of having some cages replaced with new ones, which is quite welcome given how deteriorated many of them are. Some have latches that are bent, making the doors difficult to open and close.

Although I have looked inside before, I made my first entrance into the bunny nursery today. I was reluctant to, given the prevalence of rabies in rabbits and my unvaccinated status, but an intern told me she thought the restrictions only applied to handling rabbits. For some reason, the squirrels with mange, which were previously kept in the now empty waterfowl nursery, were moved to the bunny nursery. I ended up having another crew member feed them, as this would have meant not being able to work with any other squirrels that day. I would have perhaps welcomed this automatic "out" as a non-crew leader but felt that I should not make myself unavailable in case I was needed during the rest of the shift.

25.

I would characterize today as a shift to be forgotten, if feeling incompetent plays, or ought to play, a strong role in causing forgetfulness. The reality is, however, that I learned some valuable insights in uncomfortable ways. Ironically, what I wish to forget is precisely what should not be forgotten, unless the feelings themselves can be abstracted away from the situation and the bare facts left. In general, I just felt out of sorts today. I had a hard time focusing and processing what was going on around me. It is tempting to blame this on the fact that I had to get up before 5:00 this morning to drive my grandmother to the airport, and I think tiredness played a real part in the shift being rather unsavory.

From my recollection, the difficulty began when I stuck myself for the first time with a needle. Luckily, I did not bleed this time. However, not long after this first incident I stuck myself again, and this time I bled profusely. Remembering where the first aid kit is kept in the treatment room, I disinfected the wound with an alcohol pad and placed an adhesive bandage on my finger (I learned afterward that I should have logged this in an incident report, which I do not recall having previously heard about).

After this, while drawing up meds, I absentmindedly drew several syringes far too full, having misinterpreted the hundredths as the tenths place. Luckily, my co-crew leader caught the mistake, and questioned by asking, "Who are *those* for?" Not remembering the case numbers, I replied, "Squirrels," after which a crew member quipped, "You'd make a great detective!" This incident put me in a spiral of second-guessing that lasted the rest of the shift. I felt powerless, knowledge-less, unable to make decisions. I felt as though every comment and question directed at me now was an attempt at domination, all the while trying to remind myself to not take things personally and wondering how accurate my perception was of these messages as motivated by distrust or a desire to discredit.

To make things yet more difficult, it seemed that a disproportionate number of situations I experienced today were entirely new to me. Squirrels were strangely missing from the nursery, and I didn't know how to find them. The intern was asking me pointed questions that I couldn't understand or respond adequately to. I couldn't find the squirrel with mange to be given meds because the wrong basket number was written on its med sheet—or was it? Squirrels were listed on their cage card as "MIA" (Missing in action? Were we in a war zone?). I didn't know why so many reds appeared to have thinning fur (I was told it could be an adverse reaction to meds? Or perhaps mange?). I couldn't tell if a squirrel was bloated or just fat. I didn't know if ".02 benebac" written on a food sheet meant it still needed to be given, or already had been. I didn't even know how to give benebac. Everything was framed as negative and attributed to myself by me.

Of course, these issues were not all my fault, but they certainly added to my overall feeling of disorientation. They also illustrated well the uncertainty inherent in situations and settings permeated by not only many and frequent human interactions, but also interactions with largely unfamiliar nonhuman animals and technologies. Although this uncertainty would likely subside a bit with experience, even the most experienced worker cannot possibly account for all the new and emergent variables constantly coming to be in such a setting, including new volunteers and interns, new patients and the illnesses and injuries they acquire, new medications and concentrations of medications, new policies and procedures communicated from the top down, etc. While the goal of any organization is to run smoothly, knowledge is necessary to attain this efficiency. Problematically, systems themselves do not know, but certain "nodes" within the system know some things about it. The paradox of such a setting as the WRC is that, although it is impossible to know everything relevant about the setting at any time slice, this does not absolve one of responsibility for trying to be not only omniscient, but also a perfectly controlled agent.

26.

This week went much better than the last. It seems, as I had hoped, that I learned some things of value from last week's unpleasantness. For example, I have come to believe that the incident with overdrawing meds was caused not only by tiredness, but also by my poor choice of med sheet placement. Because the area where meds are normally drawn up was cluttered, I chose to place the sheets where I could not simultaneously read them and draw up the meds. This made it necessary to try to remember the precise amount to be given, while traveling over to the med station to draw this amount. Apparently, the error of misinterpreting, for example, .04 as .4 was made on the short trip from pages to meds. Realizing that this was poor protocol, I changed to the more efficient and accurate method of keeping the sheets in front of me (and organizing the med station to make this possible, if necessary) before the ineffective approach became habitual—as had been the case with taking both clips off every basket, thereby allowing squirrels to escape.

Tuesday, I overheard an interesting discussion among the co-crew leaders about their largely negative perceptions of the new mammal nursery leader, as well as overall shortcomings in release practice. One was vociferously stating her opposition to the practice of releasing squirrels late in the day, late in the season. According to her, two that had been released were found frozen to death, and the rest were readmitted with hypothermia. She brought up some good points, among them that squirrels (especially this late in the year) should be released early in the day and with some food, since they will likely need to either make or find an unoccupied nest before retiring for the night. Since nest building in the cold takes a lot of energy, and the squirrels are young, inexperienced, and unfamiliar with their new environment, it is unlikely that they will find sufficient food in this early period of independence.

Some other observations of note from the week: Last weekend, I had the pleasure of driving up to and staying at my grandparents' cabin with my brother, Ian. The cabin is located on the South Kawishiwi River, Stony River Township, Lake County, near Ely, Minnesota. In addition to visiting the outstanding International Wolf Center (which was founded, incidentally, by L. David Mech) for the first time, I was able to observe flying squirrels for what I believe to have been the first time, as well. Due to their nocturnal habits, they are rarely seen, and although some are treated at the WRC, I have yet to see one there. Initially, I was unsure whether the squirrel I noticed at the feeder around 9:00 PM was a late-feeding red or flying squirrel, as it looked to be about the size of a red. It was dark outside and difficult to see, but after shining a flashlight on the critter I noticed it had large, beady eyes. After seeing its flattened body slide into the feeder, which the rounder reds seemed unable to do, I knew it was a "flyer."

My grandfather, grandmother, brother and I also saw an animal along the river shore which we were unable to positively identify, but which was almost certainly either an American marten (*Martes americana*) or fisher (*Martes pennanti*). Martens were nearly extirpated from Minnesota by the early twentieth century and are a great conservation success story. Although this animal seemed to be more marten than fisher-sized, it had a dark coat and looked to have smaller ears than would be expected for a marten. In addition to the trip up north, I observed in our neighborhood what may have been a white fox squirrel. It looked large enough and had an orange-ish belly like fox squirrels tend to have, but I was not positive that it wasn't a gray because of its relatively large ears. I also saw a white squirrel on campus today for the first time, reinforcing my inclination to start a University of Minnesota chapter of the Albino Squirrel Preservation Society (although the squirrel's status as a true albino remains in question).

27.

Today was my final day as crew leader in 2012, and my first shift as sole leader of the "crew," which amounted to a single volunteer. Since the season has been winding down for about a month now, my co-crew leader and I decided that there only need be one of us present at each of the last two shifts (she at last week's, and I at this week's). I was half expecting the nursery to be empty today but was somewhat surprised to see a dozen or so squirrels upon entry. Some squirrels were very large and probably ready to be released—one was around 300 grams, which I believe is the heaviest I've seen yet. Despite its size, it was quite calm and not at all aggressive. It was evident that much of the year-end cleaning had already taken place, with many objects in unfamiliar places.

I only had to give meds to two squirrels today, thankfully, with no injections. I noticed some discrepancies on the food sheets for red squirrels in outdoor cages and brought this to the attention of the nursery director, who said that the list would be adjusted. There were a surprising number of mange squirrels still being kept in the rabbit nursery, which the same intern who had led my first shift in the waterfowl nursery, fed today.

The outdoor cages are quite full now, including the two smaller rolling cages. I had forgotten the procedure, from the first half of the year, of rolling these cages inside before opening the door to put food in. This resulted in a very agile red escaping outdoors—a recipe for disaster. As the deck area surrounding the building is expansive and littered with objects a squirrel could easily hide in or under, I did not expect to recapture it and could only think what an unfortunate end to the season this escape would be. To make matters worse, the squirrel was not accustomed to the slick synthetic material the deck is made of, and upon trying to climb slipped and fell. It would have almost certainly fallen to the asphalt some three meters below, if it had not been caught by the gutter. To my astonishment, after persistently tracking the squirrel back and forth along the deck and keeping in visual contact with it for several minutes, a volunteer and I "tag teamed" with nets and recaptured it.

An email was sent out today regarding the need for two cages of squirrels to be released this weekend. Since it is late in the season and the temperatures are getting low, whoever gets to release them (on a first-come basis) also gets a nest box and food supply for the winter. This amounts to four nest boxes, each with three or four squirrel residents. Talk about deluxe accommodations! I would love to have this responsibility, but unfortunately there are too many dogs around and not enough space where I live. Hopefully the winter is not too harsh this year, and the late-released squirrels fare well.

References

Aitken, G. 2004. *A New Approach to Conservation: The Importance of the Individual through Wildlife Rehabilitation*. Aldershot: Ashgate Publishing.

Allen, T. F. H., and T. W. Hoekstra. 1992. *Toward a Unified Ecology*. New York: Columbia University Press.

Ariew, A. 2008. Population Thinking. In *The Oxford Handbook of Philosophy of Biology*, ed. M. Ruse. Oxford: Oxford University Press.

Aristotle. 1941. *The Basic Works of Aristotle*. New York: Random House.

Audi, R., ed. 1999. *The Cambridge Dictionary of Philosophy*. 2nd ed. Cambridge: Cambridge University Press.

Barsam, A. P. 2008. *Reverence for Life: Albert Schweitzer's Great Contribution to Ethical Thought*. Oxford: Oxford University Press.

Beaver, P., ed. 1982. *Wildlife Rehabilitation: Volume 1*. Smithtown: Exposition Press.

Bekoff, M. 2007. *The Emotional Lives of Animals: A Leading Scientist Explores Animal Joy, Sorrow, and Empathy—And Why They Matter*. Novato: New World Library.

Bekoff, M. 2010. *The Animal Manifesto: Six Reasons for Expanding Our Compassion Footprint*. Novato: New World Library.

Bekoff, M., and C. A. Meaney, eds. 1998. *Encyclopedia of Animal Rights and Animal Welfare*. Westport: Greenwood Press.

Benz-Schwarzburg, J., and A. Knight. 2011. Cognitive Relatives Yet Moral Strangers? *Journal of Animal Ethics* 1 (1): 9–36.

Blackburn, S. 1994. *The Oxford Dictionary of Philosophy*. Oxford: Oxford University Press.

L. B. DeVaney, *Animal Conservation Ethics and the Population Problem*, The Palgrave Macmillan Animal Ethics Series,
https://doi.org/10.1007/978-3-031-91891-9

Bruskotter, J. T., R. H. Schmidt, and T. L. Teel. 2007. Are Attitudes Toward Wolves Changing? A Case Study in Utah. *Biological Conservation* 139:211–218.

Bryant, A., and K. Charmaz. 2007. Grounded Theory in Historical Perspective: An Epistemological Account. In *The SAGE Handbook of Grounded Theory*, ed. A. Bryant and K. Charmaz. London: SAGE Publications Ltd.

Callicott, J. B. 1980. Animal Liberation: A Triangular Affair. *Environmental Ethics* 2:311–338.

Callicott, J. B. 1984. Non-anthropocentric Value Theory and Environmental Ethics. *American Philosophical Quarterly* 21 (4): 299–309.

Callicott, J. B. 1988. Animal Liberation and Environmental Ethics: Back Together Again. *Between the Species* 4:163–169.

Callicott, J. B. 1989. *In Defense of the Land Ethic: Essays in Environmental Philosophy*. Albany: State University of New York Press.

Callicott, J. B. 1996. Do Deconstructive Ecology and Sociobiology Undermine Leopold's Land Ethic? *Environmental Ethics* 18 (4): 353–372.

Callicott, J. B. 1998. 'Back Together Again' Again. *Environmental Values* 7:461–475.

Carson, R. 1962. *Silent Spring*. Boston: Houghton Mifflin.

Carthey, A. J., and P. B. Banks. 2012. When Does an Alien Become a Native Species? A Vulnerable Native Mammal Recognizes and Responds to Its Long-Term Alien Predator. *PloS One* 7 (2): e31804.

Cochran, W. W., and R. D. Lord Jr. 1963. A Radio-Tracking System for Wild Animals. *Journal of Wildlife Management* 27:9–24.

Cohn, P., ed. 1999. *Ethics and Wildlife*. Lewiston: The Edwin Mellen Press.

Crook, R. J. 2021. Behavioral and Neurophysiological Evidence Suggests Affective Pain Experience In Octopus. *Iscience* 24:3.

Daiute, C., and C. Lightfoot. 2004. *Narrative Analysis*. Thousand Oaks: SAGE Publications, Inc.

Darwin, C. 1859. *On the Origin of Species by Means of Natural Selection, or The Preservation of Favoured Races in the Struggle for Life*. London: John Murray.

Dexter, M. 2005. *Trapping Harvest Statistics*. Saint Paul: Minnesota Department of Natural Resources, Division of Fish and Wildlife.

Di Silvestro, R. L. 1985. The Federal Animal Damage Control Program. In *Audubon Wildlife Report*, ed. R. L. Di Silvestro, 130–158. New York: National Audubon Society.

Diamond, J. 1989. Overview of Recent Extinctions. In *Conservation for the Twenty-first Century*, ed. D. Western and M. Pearl. New York: Oxford University Press.

Dmytryk, R. 2012. *Wildlife Search and Rescue: A Guide for First Responders*. West Sussex: Wiley-Blackwell.

Donaldson, S., and W. Kymlicka. 2011. *Zoopolis: A Political Theory of Animal Rights*. Spain: Oxford University Press.

Elton, C. 1946. Competition and the Structure of Ecological Communities. *The Journal of Animal Ecology* 15 (1): 54–68.

Erb, J., and C. Humpal. 2020. *Minnesota Wolf Population Update 2020.* St. Paul: MN DNR.

Fuller, T. K. 1989. Population Dynamics of Wolves in North-Central Minnesota. *Wildlife Monograph* 105:1–41.

Ghiselin, M. T. 1974. A Radical Solution to the Species Problem. *Systematic Zoology* 23 (4): 536–544.

Gobster, P. H., and R. B. Hull, eds. 2000. *Restoring Nature: Perspectives from the Social Sciences and Humanities.* Washington, DC: Island Press.

Goudge, T. A. 1955. What Is a Population? *Philosophy of Science* 22 (4): 272–279.

Gould, S. J. 1990. *Wonderful Life: The Burgess Shale and the Nature of History.* New York: W. W. Norton & Company.

Gustafson, A. C. 2024. Creature Care. *Minnesota Conservation Volunteer* 87 (516): 20–31.

Heister, A. 2022. *Beyond the North American Model of Wildlife Conservation: From Lethal to Compassionate Conservation.* Cham: Palgrave Macmillan.

Hentz, P. S. 2009. *Rescuing Wildlife: A Guide to Helping Injured and Orphaned Animals.* Mechanicsburg: Stackpole.

Hindmarch, I. 1972. Observer Participation. *Drugs and Society* 1:6–10.

Horta, O. 2017. Animal Suffering in Nature: The Case for Intervention. *Environmental Ethics* 39:261–279.

Hume, D. 1739. *A Treatise of Human Nature: Being an Attempt to Introduce the Experimental Method of Reasoning into Moral Subjects.* London: John Noon.

International Union for Conservation of Nature and Natural Resources. 2025. IUCN Red List of Threatened Species, Version 2025-1. *IUCN Red List of Threatened Species.* https://www.iucnredlist.org/en. Accessed 14 June 2025.

International Wildlife Rehabilitation Council. *History.* History—International Wildlife Rehabilitation Council. https://www.theiwrc.org/history/. Accessed 11 December 2024.

Jorgensen, D. L. 1989. *Participant Observation: A Methodology for Human Studies.* Thousand Oaks: Sage.

Kant, I. 1993. *Grounding for the Metaphysics of Morals: With, On a Supposed Right to Lie Because of Philanthropic Concerns* (3rd ed.). Trans. J. W. Ellington. Indianapolis: Hackett Publishing.

Keith, L. B. 1983. Population Dynamics of Wolves. In *Wolves in Canada and Alaska*, ed. L. N. Carbyn. Ottawa: Canadian Wildlife Service. Report Series 45.

Kheel, M. L. 2000. *An Ecofeminist Critique of Holist Nature Ethics: Attending to Non-human Animals.* Berkeley: Graduate Theological Union.

Kidd, A. H., R. M. Kidd, and R. L. Zasloff. 1996. Characteristics and Motives of Volunteers in Wildlife Rehabilitation. *Psychological Reports* 79:227–234.

Lakoff, G., and M. Johnson. 1980. *Metaphors We Live By.* Chicago: University of Chicago Press.

Lederman, J. 1997. *Cries of the Wild: A Wildlife Rehabilitator's Journal.* Surrey: Heritage House.

Leopold, A. 1949. *A Sand County Almanac and Sketches Here and There.* New York: Oxford University Press.

Licht, D. S., and S. H. Fritts. 1994. Gray Wolf (*Canis lupus*) Occurrences in the Dakotas. *American Midland Naturalist* 132 (1): 74–81.

Linzey, A. 2009. *Why Animal Suffering Matters: Philosophy, Theology, and Practical Ethics.* Oxford: Oxford University Press.

Lockwood, J. A., and A. V. Latchininsky. 2008. Philosophical Justifications for the Extirpation of Non-indigenous Species: The Case of the Grasshopper *Schistocerca nitens* (Orthoptera) on the Island of Nihoa, Hawaii. *Journal of Insect Conservation* 12 (3–4): 429–445.

Lovelock, J. 2000. *Gaia: A New Look at Life on Earth.* 3rd ed. Oxford: Oxford University Press.

Lynn, W. S., L. Baker, W. T. Borrie, A. P. A. Cardilini, S. M. Alexander, S. Coghlan, P. Cryer, G. T. Bonsen, T. T. Derham, O. Keynan, C. M. Reed, S. Riley, E. A. Ryan, F. J. Santiago-Avila, K. Walker, A. E. Weiss, and N. Xenakis. 2025. Compassionate Conservation in Practice: A Values-Driven, Interdisciplinary, Pluralistic, and Deliberative Community. *Biological Conservation* 303:111002.

Massenberg, J. R., J. Schiller, and C. Schröter-Schlaack. 2023. Towards a Holistic Approach to Rewilding in Cultural Landscapes. *People and Nature* 5 (1): 45–56.

McKibben, B. 1989. *The End of Nature.* New York: Random House.

McKinney, T., S. Waters, and M. A. Rodrigues. 2023. *Primates in Anthropogenic Landscapes: Exploring Primate Behavioural Flexibility Across Human Contexts.* Cham: Springer.

Mech, L. D. 1970. *The Wolf: The Ecology and Behavior of an Endangered Species.* Minneapolis: University of Minnesota Press.

Mech, L. D. 1977. Wolf-Pack Buffer Zones as Prey Reservoirs. *Science* 198 (4314): 320–321.

Mech, L. D. 1995. The Challenge and Opportunity of Recovering Wolf Populations. *Conservation Biology* 9 (2): 270–278.

Mech, L. D. 2012. Is Science in Danger of Sanctifying the Wolf? *Biological Conservation* 150:143–149.

Mech, L. D., and R. O. Peterson. 2003. Wolf-Prey Relations. In *Wolves: Behavior, Ecology, and Conservation*, ed. L. D. Mech and L. Boitani. Chicago: University of Chicago Press.

Mech, L. D., S. H. Fritts, and W. J. Paul. 1988. Relationship between Winter Severity and Wolf Depredation on Domestic Animals in Minnesota. *Wildlife Society Bulletin* 16:269–272.

Meffe, G. K., and C. R. Carroll. 1997. *Principles of Conservation Biology.* Sunderland: Sinauer.

Menon, V., N. V. K. Ashraf, P. Panda, and K. Mainkar, eds. 2005. *Back to the Wild: Studies in Wildlife Rehabilitation*, Conservation Reference Series No. 2. New Delhi: Wildlife Trust of India.

Meriggi, A., and S. Lovari. 1996. A Review of Wolf Predation in Southern Europe: Does the Wolf Prefer Wild Prey to Livestock? *Journal of Applied Ecology* 33:1561–1571.

Merriam-Webster Dictionary. 2024. *Conservation Definition & Meaning—Merriam-Webster.* https://www.merriam-webster.com/dictionary/conservation. Accessed 8 December 2024.

Midgley, M. 1983. *Animals and Why They Matter.* Athens: University of Georgia Press.

Mill, J. S. 1864. *Utilitarianism: Reprinted from "Fraser's Magazine".* 2nd ed. London: Longman, Green, Longman, Roberts, and Green.

Miller, J. G. 1978. *Living Systems.* New York: McGraw-Hill.

Minnesota Department of Natural Resources. 2012. *2012–2013 Wolf Seasons Regulations Handbook.* Saint Paul: MN DNR.

Minnesota Department of Natural Resources. 2024b. *Wolf Management | Minnesota DNR.* https://www.dnr.state.mn.us/wolves/index.html. Accessed 9 December 2024.

Minot Daily News. 2021. Minot Zoo Veterinarian ND's First Certified Wildlife Rehabilitator | News, Sports, Jobs—*Minot Daily News.* https://www.minotdailynews.com/life/arts-entertainment/2021/06/minot-zoo-veterinarian-nds-first-certified-wildlife-rehabilitator/. Accessed 11 December 2024.

Mladenoff, D. J., T. A. Sickley, R. G. Haight, and A. P. Wydeven. 1995. A Regional Landscape Analysis and Prediction of Favorable Gray Wolf Habitat in the Northern Great Lakes Region. *Conservation Biology* 9 (2): 279–294.

Mladonicky, J., M. Sirolli, M. Willette and J. Ponder. 2024. *U.S. Fish and Wildlife Service Migratory Bird Program.* Evaluation of Current Scientific Literature on the Impact of Wildlife Rehabilitation on Conservation, Documentation of Knowledge Gaps and Identification of Future Opportunities. https://www.fws.gov/media/evaluation-current-scientific-literature-impact-wildlife-rehabilitation-conservation. Accessed 1 March 2025.

Moellendorf, D. 2020. Three Interpretations of the Anthropocene: Hope and Anxiety at the End of Nature. *Ethics, Politics & Society* 3:133–153.

Moore, G. E. 1993. *Principia Ethica.* 2nd ed. Cambridge: Cambridge University Press.

Naess, A. 1973. The Shallow and the Deep, Long-Range Ecology Movement: A Summary. *Inquiry* 16:95–100.

Naess, A., and I. Mysterud. 1987. Philosophy of Wolf Policies I: General Principles and Preliminary Exploration of Selected Norms. *Conservation Biology* 1 (1): 22–34.

Nagy, K., and P. D. Johnson II, eds. 2013. *Trash Animals: How We Live with Nature's Filthy, Feral, Invasive, and Unwanted Species.* Minneapolis: University of Minnesota Press.

National Wildlife Rehabilitators Association. 2024. *About Us.* National Wildlife Rehabilitators Association. https://www.nwrawildlife.org/page/AboutNWRA. Accessed 11 December 2024.

National Wildlife Rehabilitators Association and International Wildlife Rehabilitation Council. 2021. *Standards for Wildlife Rehabilitation.*

North Dakota Game and Fish Department. 2024. Orphaned and Injured Animals | North Dakota Game and Fish. Accessed 11 December 2024.

Oxford English Dictionary. 2024. Biology—Quick search results | Oxford English Dictionary. https://www.oed.com/search/dictionary/?scope=Entries&q=biology. Accessed 8 December 2024.

Peggs, K. 2012. *Animals and Sociology.* Hampshire: Palgrave Macmillan.

Pesic, P. 1999. Wrestling with Proteus: Francis Bacon and the 'Torture' of Nature. *Isis* 90 (1): 81–94.

Pyke, G. H., and J. K. Szabo. 2018. Conservation and the 4 Rs, Which Are Rescue, Rehabilitation, Release, and Research. *Conservation Biology* 32 (1): 50–59.

Regan, T. 1983. *The Case for Animal Rights.* Berkeley: University of California Press.

Rojas, M. 1992. The Species Problem and Conservation: What Are We Protecting? *Conservation Biology* 6 (2): 170–178.

Rolston, H., III. 1989. Biology without Conservation: An Environmental Misfit and Contradiction in Terms. In *Conservation for the Twenty-first Century*, ed. D. Western and M. C. Pearl. New York: Oxford University Press.

Ross, L., and R. E. Nisbett. 1991. *The Person and the Situation: Perspectives of Social Psychology.* New York: McGraw-Hill.

Rud, A. G. 2011. *Albert Schweitzer's Legacy for Education: Reverence for Life.* New York: Palgrave Macmillan.

Ruse, M. 1990. Are Pictures Really Necessary? The Case of Sewell Wright's 'Adaptive Landscapes'. *PSA: Proceedings of the Biennial Meeting of the Philosophy of Science Association* 2:63–77.

Ryder, R. D. 1991. *Painism: Ethics, Animal Rights, and Environmentalism.* Cardiff: University of Wales.

Ryle, G. 1949. *The Concept of Mind.* London: Hutchinson House.

Sandler, R. 2010. The Value of Species and the Ethical Foundations of Assisted Colonization. *Conservation Biology* 24 (2): 424–431.

Sandler, R. 2012. Intrinsic Value, Ecology, and Conservation. *Nature Education Knowledge* 3 (10): 4.

Schlaepfer, M. A., and J. J. Lawler. 2023. Conserving Biodiversity in the Face of Rapid Climate Change Requires A Shift In Priorities. *Wiley Interdisciplinary Reviews: Climate Change* 14 (1): e798.

Schlaepfer, M. A., D. F. Sax, and J. D. Olden. 2011. The Potential Conservation Value of Non-Native Species. *Conservation Biology* 25 (3): 428–437.

Schweitzer, A. 1950a. In *The Animal World of Albert Schweitzer: Jungle Insights into Reverence for Life*, ed. C. R. Joy. Boston: Beacon Press.

Schweitzer, A. 1950b. *The Philosophy of Civilization.* Trans. C.T. Campion. New York: Macmillan.

Schweitzer, A. 1969. *Reverence for Life.* Trans. R.H. Fuller. New York: Harper & Row.

Schweitzer, A. 1992. *Letters 1905–1965*. Trans. J. Neugroschel. New York: Macmillan.

Simberloff, D., and T. Dayan. 1991. The Guild Concept and the Structure of Ecological Communities. *Annual Review of Ecology and Systematics* 22:115–143.

Singer, P. 1975. *Animal Liberation*. New York: Avon.

Slingerland, E. G. 2008. *What Science Offers the Humanities: Integrating Body and Culture*. Cambridge: Cambridge University Press.

Smith, D. 2012a. It's open season on rules for Minnesota wolf hunt. *Star Tribune. May* 22:2012.

Smith, D. 2012b. Minnesota DNR Survey: Most Oppose Wolf Hunt. *Star Tribune*, June 22, 2012.

Sneddon, L. U. 2020. Can Fish Experience Pain? In *The Welfare of Fish*, Animal Welfare 20, ed. T. Kristiansen, A. Fernö, M. Pavlidis, and H. van de Vis. Cham: Springer.

Soule, M. E. 1985. What Is Conservation Biology? *BioScience* 35 (11): 727–734.

Soule, M. E., ed. 1987. *Viable Populations for Conservation*. Cambridge: Cambridge University Press.

Stearns, F. 1990. Forest History and Management in the Northern Midwest. In *Management of Dynamic Ecosystems*, ed. J. M. Sweeney. West Lafayette: The Wildlife Society, North Central Section.

Steinfeld, H., P. Gerber, T. Wassenaar, V. Castel, M. Rosales, and C. de Haan. 2006. *Livestock's Long Shadow: Environmental Issues and Options*. Rome: United Nations, Food and Agricultural Organization.

Stocking, G. W. 1960. Franz Boas and the founding of the American Anthropological Association. *American Anthropologist* 62:1–17.

Thain, M., and M. Hickman. 2004. *Penguin Dictionary of Biology*. 11th ed. London: Penguin Books.

Thiel, R. P. 1985. Relationship between Road Densities and Wolf Habitat Suitability in Wisconsin. *American Midland Naturalist* 113:404–407.

Thompson, B. H., Jr. 1997. The Endangered Species Act: A Case Study in Takings and Incentives. *Stanford Law Review* 49:305–380.

Treves, A., L. Naughton-Treves, E. K. Harper, D. J. Mladenoff, R. A. Rose, T. A. Sickley, and A. P. Wydeven. 2003. Predicting Human-Carnivore Conflict: A Spatial Model Derived from 25 Years of Data on Wolf Predation on Livestock. *Conservation Biology* 18 (1): 114–125.

Treves, A., R. L. Jurewicz, L. Naughton-Treves, and D. S. Wilcove. 2009. The Price of Tolerance: Wolf Damage Payments after Recovery. *Biodiversity Conservation* 18:4003–4021.

U.S. Department of the Interior. 2011. Endangered and Threatened Wildlife and Plants; Revising the Listing of the Gray Wolf (*Canis lupus*) in the Western Great Lakes. *Federal Register* 76(249):81666–81726. https://www.federalregister.gov/documents/2011/12/28/2011-32825/endangered-and-threatened-wildlife-and-plants-revising-the-listing-of-the-gray-wolf-canis-lupus-in. Accessed 8 December 2024.

U.S. Fish and Wildlife Service. 2006. *Gray Wolf Biology Questions and Answers.* Washington, DC: USFWS. https://www.fws.gov/sites/default/files/documents/2024-04/1.pdf. Accessed 8 December 2024

U.S. Fish and Wildlife Service. 2011. *Gray Wolf Recovery in Minnesota, Wisconsin, and Michigan*. Washington, DC: USFWS. https://wisconsinwolffacts.com/forms/fedwolfplan.pdf. Accessed 8 December 2024.

U.S. Fish and Wildlife Service. 2023. Washington, DC: USFWS. *USFWS Gray Wolf Statement.* https://www.fws.gov/sites/default/files/documents/2023%20USFWS%20Gray%20Wolf%20Statement.pdf. Accessed 8 December 2024.

United Nations Department of Economic and Social Affairs. 1968. *The Concept of a Stable Population: Application to the Study of Populations of Countries with Incomplete Demographic Statistics.* New York: United Nations.

Von Uexkull, J. 1920. *Umwelt und Imnenwelt der Tiere.* 2nd ed. Berlin: Springer.

Vors, L. S., J. A. Schaefer, B. A. Pond, A. R. Rodgers, and B. R. Patterson. 2007. Woodland Caribou Extirpation and Anthropogenic Landscape Disturbance in Ontario. *Journal of Wildlife Management* 71 (4): 1249–1256.

Waldau, P. 2010. *Animal Rights: What Everyone Needs to Know*. New York: Oxford University Press.

Wallach, A. D., C. Batavia, M. Bekoff, S. Alexander, L. Baker, D. Ben-Ami, L. Boronyak, A. P. A. Cardilin, Y. Carmel, D. Celermajer, S. Coghlan, Y. Dahdal, J. J. Gomez, G. Kaplan, O. Keynan, A. Khalilieh, H. Kopnina, W. S. Lynn, Y. Narayanan, S. Riley, F. J. Santiago-Ávila, E. Yanco, M. A. Zemanova, and D. Ramp. 2020. Recognizing Animal Personhood in Compassionate Conservation. *Conservation Biology* 34:1097–1106.

Warren, K. J. 2015. Feminist Environmental Philosophy. In *Stanford Encyclopedia of Philosophy.* https://plato.stanford.edu/entries/feminism-environmental/. Accessed 10 December 2024.

Wilcove, D. S., and M. Wikelski. 2008. Going, Going, Gone: Is Animal Migration Disappearing? *PLoS Biology* 6 (7): e188.

Williams, J. 2022. 2021 was the Biggest Year Ever for Wildlife Rehab at Minnesota Center. *Star Tribune.*

Wittgenstein, L. 1958. *Philosophical Investigations.* Trans. G.E.M. Anscombe. Oxford: Basil Blackwell.

Wydeven, A. P. 1994. Travels of a Midwestern Disperser. *International Wolf* 4 (1): 20–22.

Young, S. P., and E. A. Goldman, eds. 1944. *The Wolves of North America.* New York: Dover Publishers.

Zs.-Nagy, I. 1994. *The Membrane Hypothesis of Aging.* Boca Raton: CRC Press.

Index[1]

[1] Note: Page numbers followed by 'n' refer to notes.

L. B. DeVaney, *Animal Conservation Ethics and the Population Problem*, The Palgrave Macmillan Animal Ethics Series, https://doi.org/10.1007/978-3-031-91891-9

V

W